A Life and Career in Chemistry

Pierre Laszlo

A Life and Career in Chemistry

Autobiography from the 1960s to the 1990s

 Springer

Pierre Laszlo
Sénergues, France

ISBN 978-3-030-82395-5   ISBN 978-3-030-82393-1   (eBook)
https://doi.org/10.1007/978-3-030-82393-1

This Springer imprint is published by the registered company Springer Nature Switzerland AG.
The registered company address is: Gewerbestrasse 11, 6330 Cham, Switzerland

Introduction

Amateurs of baroque music keenly hear the difference between an ancient harpsichord and a modern instrument, even when the latter is a reconstitution. Richness of tone makes the difference between the old and the new. Why are we, the self-labeled sophisticate moderns, unable to match the prowess of yesteryear? In short, because the tone of an instrument is a multiparameter feat. Harpsichords are based on plucking strings. A plectrum activates each individual string. The plucking submits to a wealth of parameters: in addition to strong or weak, the plectrum can hold on to the string, roll, twist and turn, bite, scratch, whip it, all variants that contribute to the tone.

The analogy is to the historian, whose position is akin to that of the modern instrument builder. He or she is unable to recapture the rich tone of the past, in its many-stranded fabric. The task resembles hauling water with a basket, to use a cruder metaphor.

Far from an ego-boosting adventure, this book is meant primarily for future historians of twentieth-century chemistry. They will be able to peruse it, not necessarily for the ostensible information about my times, more efficiently for implicit and revealing tidbits. And that is the value of this unjustly decried genre, the autobiography.

I submit mine in a spirit of modesty, not self-deprecation. I wrote it because the reading of autobiographies of scientists was an important part of my education and of espousing scientific research as my career, instead of other endeavors that were equally tempting, literature foremost. Which autobiographies?

I will cite only these: Benjamin Franklin's, Pierre-Gilles de Gennes's, Benoît Mandelbrot's, Laurent Schwartz's, and Jacques Friedel's. From each, I learned the virtue of being different and forging ahead.

In addition, prior to engaging in this exercise in frankness and memorization, I worked on portrait-drawing in words. At the time of writing, the early summer of 2020, I have penned and published the portraits of nearly 70 alumni of the French École polytechnique, all of whom had their training in the sciences and some of whom became scientists themselves.

How then can a single person help to preserve bits and pieces of the past? By contributing a tone of voice, maybe. Not shying away from one's singularity. Describing one's sights, encounters, and experiences.

Thus, I submit that personal histories may guide science history. By heeding such an axiomatic precept, I feel very much a product of my time, of the sixties when I became a member of the scientific community—a notion also from that time and worth reexamining. An injunction from that period was to make a contribution. Which is exactly what this book will strive for.

A feature of the sixties appealed to me and thus may feature predominantly in this memoir. An epistemology of combinatorials, which applied to both linguistics and chemistry—I will only mention at this point my published suggestion to teach chemistry as a language, which would help to tackle the exponential accumulation of chemical knowledge during the last century.

Why bother writing an autobiography? Assuredly not for self-glorification: science has lavished on me all kinds of rewards; I do not have a need to add to them. A more lasting note is to pass on the experience of rising to a variety of challenges. My take on the autobiographical foray is to stress the personal, in my case the permanent tension between science and the humanities, between chemistry and literary studies, between Hephaistos and Athena—to put it under the aegis of Greek gods. Let me note in passing, a point I probably will not have the time and space to elaborate further, the absurdity of keeping separate histories of art and of science. Regarding the latter, I witnessed the changing of the guard: a switch rather than a mere shift. In the past, historians of science had a dual training in science and in history. Double doctorates were not unusual. Such an exacting training has now been jettisoned. History of science has undergone a takeover by sociologists, some of whom are dropouts from scientific studies, even sometimes scientiphobes who blame science on the fallout from technology—the term technoscience is revealing—thus putting in the same disposable bag two developments characteristic of the past century, advancement of knowledge and consumerism. Which is axiologically wrong, akin to blaming philosophy for the political ills of today's world.

Why do we need history of science? What is good about it? The advancement of knowledge, far from being linear (Whig historiography), is replete with twists and turns. New departures are the norm. They originate from ideas, hence from people: yes, I am an idealist; Plato had it right. To chronicle these abrupt changes is the task of the science historian. The reward is archival work, very tedious as a rule but relieved by occasional bursts of life, which suddenly spring at you from yellowed documents.

Contents

Family and Upbringing

1

I am the son of immigrants. I was born in Algiers in August 1938. Both my parents were Hungarian.

My father, François Laszlo (1907–1976) (in Hungarian, László Ferenc; in like manner to many among his contemporaries thus named in honor of Franz-Josef, then the beloved Emperor of Austria and Hungary) attended a secondary school run by the Piarist Fathers. He received top grades in every subject. He could even speak Latin! After graduation, he attended the Technical University in Budapest—the uppermost in the country—and obtained a degree in mechanical engineering (Fig. 1.1).

Throughout life, he remained proud of being an engineer. Whatever his position in R&D or business, his calling card simply read "François Laszlo, *ingénieur*."

He complemented his degree with one year of additional training at the Politecnico, in Milano. After his return to Hungary, he was hired in a manufacture of harvesters. Because of the Depression this job lasted only for a year.

My grandparents gave him their savings to help find a job outside Hungary, then crippled economically. Algeria was his choice.

At the time, in 1930, it was a French colony. France had—maybe still has to a degree—a tradition of ignoring foreign degrees. My father was thus unable to work as an engineer. He had to work as a technician—at first to a vintner in Boufarik, the city where the softdrink Orangina was devised and first manufactured; later on as a draughtsman, in Algiers—he had to remain as such, at least formally, until after World War II, when he could return to the university in Grenoble and get a French degree (Fig. 1.2).

This was in 1946: my Dad qualified in a single year for both a BS in mathematics and an engineering degree in hydraulics. But I am getting ahead of my story.

I know less of my mother's life, she died when I was 7, giving birth to my first brother. Named Madeleine Aczel (1910–1945) (Aczel Magda in Hungarian) she was younger than my father by three years. An only child, after attending the only secondary school in Budapest for girls, she supported herself doing free-lance work as a commercial correspondent, primarily in German. Her real interest, though,

P. Laszlo, *A Life and Career in Chemistry*,
https://doi.org/10.1007/978-3-030-82393-1_1

Fig. 1.1 My father, François
Laszlo, in the Thirties. Often,
he was impeccably dressed.
Photographer unknown,
author's collection

was poetry. A devotee of Heine and Rilke in German, of Villon in French, she wrote
poems in Hungarian.

She and my father dated when he was at the University in Budapest and before he
spent a year in Italy. I do not know if they were formally engaged. In any case, as
soon as he had settled in Boufarik, she joined him there—travelling by herself across
half of Europe—she was very strong-willed. They were married a few weeks later, in
November 1932. He was 25, she was 22. Soon afterwards, they moved to Algiers
(Fig. 1.3).

Magda Laszlo then re-invented herself. Not as a French language poet, but as an
Eastern European instructor in dance, more accurately in rhythmic gymnastics. My
mother was very much influenced by the teachings of modern dancers and theorists
such as Isadora Duncan (1877–1927) and Émile Jaques-Dalcroze (1865–1950). She
opened a dance studio in our apartment in the center of Algiers. Soon, it was popular
with French young ladies of the bourgeoisie and it raised the social status of my
parents and made them a number of close friends: Pierre Goinard (1903–1981), a
brain surgeon, and his wife Henriette Goinard (1909–1991), an intellectual in like
manner to Magda (I was named after Dr. Goinard). In addition, my parents knew

Fig. 1.2 A house in Algiers, my father helped design and build. Photo by Henri Eichaker

Albert Camus (1913–1960), then a Communist and a journalist in the local newspaper *Alger Républicain*.

During the Thirties, my parents used their Central European contacts to provide intelligence to the French Army Deuxième Bureau, some officers of whom—one named Malgrat—became their friends. The information, I imagine, had to do with German war preparations.

From 1934 to 1938, François Laszlo worked for an architectural firm in Algiers. He designed a movie theater. He also filed for a patent on use of compressed reeds as a building component. Both these projects outgrew the architectural blueprints he was paid for.

Then, he changed jobs. He moved from an architectural firm to one specialized in hydraulics, specifically irrigation systems, a necessity in the hot Algerian climate. This company was a subsidiary of Neyret-Beylier-Picard-Pictet, based in Grenoble, in Continental France—a name later shortened to Neyrpic—mostly active in manufacturing turbines for hydroelectric plants.

Everything changed for our small family at the beginning of 1940, during the period known as the Phony War. The three of us moved in March 1940 from the safety of Algiers to the very relative safety of Grenoble. Why such a move, that brought us much closer to danger—and to near-starvation towards the end of the war?

The directorate at Neyrpic, a not very large family-owned company, asked this gifted young Hungarian engineer, whose salary was to them a bargain, to join the

Fig. 1.3 My mother Magda,
in Algeria in the early Thirties
(author's collection)

group of engineers in Grenoble. In addition, a conjecture on my part, the French
intelligence officers in charge of François and Magda Laszlo may have ordered them
to accept the invitation and continue their valuable intelligence work in France.

On that point, my father indeed continued during the Occupation years to provide
intelligence to his *Deuxième Bureau* contacts. Moreover, he joined the Resistance
network in Grenoble headed by a priest, known as Abbé Pierre (Henri Groués,
1912–2007), involved in exfiltrating threatened Jews, children in particular, to the
safety of neutral Switzerland. I have a recollection of witnessing the joyful relief of
my parents after my father had escorted a Monsieur Simon across the Swiss border in
Annemasse, during a week we spent in nearby Juffly-Fillinges, in the Haute-Savoie.

My mother was also politically active. Grenoble had a manufacturing plant for
artificial silk, using the Viscose process. This procedure had been first used in
Hungary. Accordingly, Hungarian workers, females predominantly, had been
imported for their familiarity with the tricky process. Magda Laszlo made it her
task to unionize them and raise their political awareness.

What do I owe to my parents, culturally in addition to genetically? A whole lot. Being hardworking, with the example of my Dad throughout his life—he died in 1976 from a combination of deep depression and an unfortunate liver-destroying medical treatment.

Being bookish, Magda especially being very well-read, I have—and do not cherish—a recollection of her reading to me Schiller (1759–1805) in the original German, I was only 5 or 6. What an ordeal! I was attentive, as she had ordered me to, but could not understand a thing.

Her favorite French writer was the pacifist Romain Rolland (1866–1944), who introduced her to the Indian poet Rabindranath Tagore (1861–1941). My parents would collect me from school, on Saturdays at 4 in the afternoon, for a cherished small ritual: first, in a bookstore where they would buy me a book, any book of my choice; and second to a pastry shop, where I could be gratified in like manner.

As already mentioned, my mother died in June 1945, when I was not quite 7. I vividly recall the harsh gymnastic exercises she was tutoring me through every morning until then. As a consequence of her death, not only were they discontinued but in addition my body underwent a silent protest, I became very stiff for the rest of my life while also inheriting the bad back both my parents suffered from!

While never saying so explicitly, my father projected to me the literary skills of my mother. In his eyes, I had inherited them. I did my best to own up to these expectations. This went with another outstanding characteristic of his personality: to him I was an individual, worthy of respect for my differences—not simply an extension of his.

My parents of course spoke Hungarian to one another—but only French to me, wanting me to be well inserted into their chosen new country. Accordingly, I never knew Hungarian and was never able to speak it fluently. However, having it heard it being spoken, with my brain unwittingly trying to make sense of the foreign words, gave me something of an ability: I am good at languages. Linguistics have always fascinated me.

My mother's untimely death—she was only 34—led to my father being an amazing parent, he somehow managed to being in his affection and care both father and mother. In one respect, her death led to a radical switch. She had raised me very much as a daughter. Blond with blue eyes, I wore curls until aged about 4. Prior to her death, I had attended *Lycée de filles*: a small group of boys, about 10, in a whole class of girls, about 30. She feared for me the brutality of boys. In October 1945, I started attending *Lycée de garçons*.

I owe to my father another asset: this is the *"François Laszlo, ingénieur"* feature of self-assertion, of not being intimidated by titles or positions of prestige. I have never shied away from addressing anyone and, conversely, I have seldom been guilty of pulling rank. Being just myself is my anchor and pedestal both.

Let me return here to my Dad lacking a French degree. In 1946, with an intervention from Abbé Pierre who had become a representative in the National Assembly, he gained French citizenship. He enrolled at the University in Grenoble and within a year earned a BS in mathematics—and was tempted to go on and become an academic—and an engineering degree in hydraulics. Thus, at long last, his take-home salary was commensurate with his contributions to the Neyrpic plant.

He then remarried, with another Hungarian lady, Fodor Ella (1913–1999). She had a daughter, six years younger than I and had lost her husband in the war. She was amazing in her generosity and warmth. She managed to build a genuine family out of scraps, so to say: her daughter Ilona (b 1944), myself and my brother Jean-François (b 1945), plus my other brother Jean-Louis, born to them in 1954.

I owe her, most of all, my love of classical music. She was a very good amateur pianist. Through her, I gained a deep-rooted admiration of Bach. In my early teen years, I vividly recall my listening repeatedly to the Wanda Landowska's (1879–1959) recordings—among the first LPs—of the "Well-Tempered Clavier" on the harpsichord. Later on, she and I communed likewise in admiration for another genius of the keyboard, Glenn Gould (1932–1982).

What my three parents taught me, in addition to love of music and literature, to not being intimidated by numbers and equations? Self-worth, as already pointed at. Hence, fierce individualism. As wrote Henri Michaux, "*Qui chante en groupe mettra lorsqu'on le lui demandera son frère en prison*" (Who sings in a group, when asked, will put his brother in jail). The power of imagination, of conceiving personal projects and following through on the idea. My father made a recommendation, followed gratefully: when you have completed a chapter of study, before turning to the next, publish a book on it. Which I am doing here, now that I have reached 80. What will the next chapter consist of, I wonder!

A Series of Great Teachers

2

Luck in life is a gift from the Gods. I am a lucky person. As part of my luck, I had outstanding teachers.

The first was Mademoiselle Guillet, in kindergarten, which I attended at *École maternelle de la Bajatière* in Grenoble for two or three months in the spring of 1944. Why was she memorable? She was utterly unconventional—I submit one of the marks of a great teacher. She taught us how to write, not with a pencil, not with a pen—the then ubiquitous *porte-plume*, pen holder that you dipped into an inkwell. She taught us to write with a paintbrush, in the manner of Chinese or Japanese calligraphers. The ability thus gained stayed with me. It is so important to write very legibly, so that going through one's lecture notes is made easier, a talent that became precious much later on, when I attended *classe préparatoire* (more on that later).

Another admirable feature of Mademoiselle Guillet was her ability at improvisation. One morning, for instance, she declared: "we won't have class today. There are *hannetons* (May beetles) in the courtyard. Come along, we'll look at them." And she launched into biology, pointing us to what to observe in those insects—now more or less extinct due to pesticides, making them a sight of the past.

After the schoolyear was over, in mid-July 1944, when I was a little short of six, my parents then entrusted me to her care for tutorials. She did it out of generosity, not for money; and taught me how to read, in less than a month. The universe of books, henceforth, was mine. My ability to read made me skip first grade, I entered *dixième* (second grade) in the fall of 1944.

I'll jump now to Monsieur Guittard, who taught the *huitième* (4th grade), in 1946–47, at Lycée Champollion, the secondary school for boys in Grenoble. In those years, all the teachers there were men. A lady might show up, exceptionally and very briefly, when there was no other option.

In any case, Guittard exuded competence and the joy of knowledge. I idolized him. He taught most of the subjects, French, arithmetic, history and geography, etc. He also introduced a light dose of science, using a *leçon de choses* book: each of the topics was given a double page, with a narrative, a picture and a summary meant to be memorized.

P. Laszlo, *A Life and Career in Chemistry*,
https://doi.org/10.1007/978-3-030-82393-1_2

There was an incident that year I was involved in. I went back home from the school shattered and sobbing. I had received a grade of zero on a test. My father went to the school to investigate. The teacher was convinced that I had cheated, since my paper was 100 percent identical with the material in the book. Fortunately, my Dad was able to convince him that it was utterly consistent with my excellent memory.

Monsieur Guittard had a highly beneficial influence on me. In the aftermath of my mother's death, no doubt as a consequence, I started stuttering badly. My father signed me up for afterschool tutorials with Monsieur Guittard. After a few such sessions, the problem disappeared. Monsieur Guittard must have told me things to boost my self-confidence.

Was I lucky again with Monsieur Francès (1904–1984), who taught mathematics in classe de *sixième* (6th grade) during the 1948–1949 year, also at Lycée Champollion. He was a phenomenon, a force of nature, a born teacher and a lover of astronomy.

He was a big man, a colossus. Grenoble is in a rugby-playing region and he belonged to a rugby team. His physical strength was legendary. Was it mythical? He had the reputation of having bodily carried out of the classroom and thrown out a miscreant pupil; I don't believe it was a tall tale.

I vividly recall indeed a ploy of his that, we the students, rejoiced in setting up. We would remove his chair from behind the desk and set it at a distance, by the door. The break over, he would reenter the classroom, grab the chair gingerly from its back, hold it horizontally with his fingers, carry it in this manner and delicately set it down again behind the desk.

But he was also a born teacher. He loved maths and he had a talent for conveying his enthusiasm for the subject. Moreover, he rejoiced in teaching fresh minds in the very first year of secondary school. I'll mention him again in this narrative.

My first inkling of chemistry came a year or two later from our biology teacher, Monsieur Bertrand. Irascible and grumpy, his one redeeming feature was his geology avocation, that he shared with his students. He must have been a devoted amateur mineralogist. When he mentioned that hobby, he became illuminated from the sheer beauty of the stones he had unearthed and studied. Later reading some of Ernst Jünger's writings, I was struck by this common vein with Bertrand's lecturing digressions. Bertrand even attempted explaining to us the chemistry involved in the colors of minerals.

Aged 12, I accompanied my father when he was sent to Brazil by the Neyrpic company—setting dams on the Rhône river had completed the French hydroelectric infrastructure. He had the mission of exploring Latin America for potential new markets. A few months later, the rest of the family followed. During the fall of 1951, I was at Lycée Pasteur in São Paulo. I completed there the *troisième* (ninth grade) in a couple of months.

Following which, after we had moved to Rio de Janeiro, I attended there the *Liceu franco-brasileiro*, in the Laranjeiras district. During our first year in Rio, I was in *première* (tenth grade). All our classes, except for the Portuguese language, were taught in French.

Some of the teachers were very good. One was exceptional, Senhor Bahiana, who taught us chemistry. For a Brazilian, his name was rather short, Henrique de Paula Bahiana.

A mention in passing: his brother-in-law, Georges Neu (1909–1995) taught us physics. Georges Neu was an alumnus of *École polytechnique*. He graduated in 1930. His younger brother Charles Neu (1914–2000) also attended École polytechnique and graduated in 1935 in the *Corps du Génie Maritime* (naval engineering). This may explain Georges's visit to Brazil, where he fell in love with and married in 1938 Maria Alicia de Cunha Bahiana. Georges Neu was an excellent physics teacher. I recall his presentation of geometrical optics, including a very didactic description of how a *cataphote* (retroreflector) works.

But I want to emphasize here my introduction to chemistry, thanks to Professor Bahiana. He taught, at Liceu Franco-Brasileiro, in a small amphitheater lecture room, outfitted with a demonstration bench and a hood.

Seating arrangement in this classroom was not optional. Professor Bahiana had the girls in the class sit in the front—starting at Lycée Pasteur in São Paulo, the classes I attended had both boys and girls—and the boys behind them. The sitting rows were separated from the rostrum by a railing.

After a couple of lectures, Professor Bahiana made me sit on the other side of the railing at a desk, smaller than his. He gave me a dual assigned task. I had to equilibrate, i.e., set the correct coefficients, the chemical equations he wrote on the blackboard. Plus I was his assistant in the demonstrations.

Which was a source of great tension (and huge fun). He instructed me as they began. He had a vantage point, the threshold to the door into the classroom: a single step would ensure his safety in the courtyard, should the worst occur.

Meanwhile, I would go on preparing a mix and transferring stuff into flasks. In doing so, I had a dilemma, the Professor was issuing instructions from his distant location and my classmates, who yearned for a gigantic pyrotechnic show, whispered to me to double the amounts in order to create a fiercer explosion!

All in all, I learned quite a bit of experimental chemistry—and I held on to all ten fingers. Professor Bahiana may have been shy about running demonstrations himself, but he was quite a wizard on practical matters: he taught us for instance how to fake sparkling wine with tartaric acid and gave us the recipe for Berthollet's powder (sugar and potassium chlorate, to which I shall return). Plus, he did not fail to teach us rudiments in the history of chemistry. Fluent in French, masterly in teaching imaginatively, this Brazilian gentleman had a powerful intuition: how else account for his choosing me to be his assistant? Truly, he was a great teacher.

During the winter of 1953, we returned to Grenoble. Reneging on his promise of a one-month vacation in Mégève, a mountain resort, my father enrolled me in January 1954 in the already started (in October) final year of secondary school, again at Lycée Champollion. Only 15, I was not set on a definite career. My parents pushed for me to become either a physician or a banker, neither of which appealed to me. The section I attended was *Sciences Expérimentales*, that prepared to medical school. Lucky guy that I am, the maths teacher was again Monsieur Francès. In early summer of 1954, I graduated from the *lycée*.

Into what, I had no idea. I was seduced by two very distinct courses, both preparing for the entrance examination to the *École normale supérieure*, either in philosophy and literature, or in the sciences. Both would entail attending a *classe préparatoire—prépa* for short—, offered at Lycée Champollion. In France, the system of *prépas* is parallel to the universities, at a higher level of difficulty. A comparison, but not wholly adequate, is with liberal arts colleges in the US.

Upon telling my Dad about both this goal and my undecidedness, he said, "let's go and talk to the head at Lycée Champollion." This person, Raymond Schiltz (1902–1984), recommended the sciences option; for which, my training in maths being too weak, I would have to do again the final grade of school, in the section known as *Mathématiques élémentaires*. It was taught, in maths, by Monsieur Francès (1904–1984)—a fact which weighed in my enthusiastic acceptance of the suggestion.

And indeed the material he taught was almost totally new to me: trigonometry and spherical trigonometry, algebra, classical geometry in both the plane and three dimensional space, derivatives and integrals, some elementary number theory, logarithms, . . .

In addition, Monsieur Francès felt free to tell us at some length of his avocation, navigation and astronomy. He taught us about the planet: longitude and latitude, their determination; time zones and the dateline; great circles and *loxodromie* (rhumb line). And he proceeded to teach also about stars and the solar system. Once again, he was just great.

After I passed again, rather easily, the graduating exam of *baccalauréat*, a normal course of study would have had me attending a *prépa* known as *Mathématiques supérieures*, also at Lycée Champollion.

It entailed studying descriptive geometry. Which in turn demanded skill in drawing for which I was utterly lacking—perhaps in reaction to my father being so good at it, after having been compelled to be a draughtsman as well as an engineer for a quarter of a century.

Once again my Dad deferred to my wishes. He pulled a stunt, finding the only *prépa* in the whole of France that lacked descriptive geometry in its cursus. It existed in Paris, at the Lycée Saint-Louis, and was known as *Normale Sciences Expérimentales*—NSE for short. Its function was to recruit future naturalists into the *École normale supérieure*. The covert—never mentioned explicitly—formula of NSE was to match the best French students with the best French science educators: no wonder if most of us (90%?) became university professors!

The encyclopaedic format of NSE consisted basically of three subjects, maths, physics/chemistry and biology. These were taught by Chazal, a Bourbaki mathematician, Privault who taught us botany and zoology, and Deluchat for physics and chemistry. All three were graduates from *École normale supérieure* (ENS). All three taught their science as it had become by the first World War, i.e., basically the big corpus the nineteenth century had assembled.

I shall delve into Deluchat's teaching because it was truly admirable, a jewel. René Deluchat entered *ENS* in 1923, from Lycée Jules Ferry in Limoges, where he captained the rugby team—an avocation he continued at the ENS. There, he came under the spell of physics, as taught by Georges Bruhat and embodied in several of his monographs; his favorite in physics was thermodynamics. He also became close friend at that time with the writer Jean Prévost (1901–1944), who had entered the ENS in 1919 and who would become a hero in the French Resistance during Occupation by the Germans.

While at the ENS, Deluchat—a workaholic—came under the spell of chemistry as well. He achieved a doctorate in that subject, preparing novel acetylenic derivatives, under the guidance of Robert Lespieau (1864–1947).

After his certification from the *agrégation* (a top level examination for teachers) he taught secondary school in several small cities, Pontivy, Vendôme and Chartres (one year in each), prior to being appointed in Paris, at the *prépa* in Lycée Buffon. He began teaching physics and chemistry at NSE in the fall of 1952.

What was remarkable about his teaching? Briefly: clarity, beauty of expression, wit and a sense of drama. He taught chemistry not from the outside as a physicist, then the norm, but from inside. In addition, he prided himself and we hugely benefited from him running the chemistry laboratory sessions. He loved handiwork and tinkering and it showed.

A word in passing about Professor Deluchat indulging an idiosyncrasy of mine. During the lab sessions, in-between experiments, I would make chemograms by mixing chemicals on a piece of filter paper and admiring the resulting color mix. I even toyed with the idea of making a career of it—this was the time of Abstract Expressionism. Deluchat knew what I was doing, it was completely foreign to the curriculum but he nevertheless indulged this fancy of mine.

I should add Madame Barberon to even a short list of teachers who strongly influenced me. She was a secondary school teacher and I served a one-month internship in her classroom at the Camille Sée Lycée in Paris. She had a forceful presence and a commanding voice: I vividly recall her commenting after my first lecture in her class, "you are like all beginners, you don't know how to set your voice." She was extremely liberal, giving me total freedom in the material I presented. Which included a practical demonstration of conservation of rotational momentum, which I explained to the students was applicable to making a skiing turn. The students at Camille Sée were only girls.

The same year 1961 as that internship, I attended the classes in mathematical methods of physics (MMP) taught by the great mathematician (Fields medalist), Laurent Schwartz (1915–2002). In addition, he was a born teacher: lively, with an arresting presence, a virtuoso in presenting most simply very complex material. My recollection of MMP is colored by a rather dramatic failure of mine. During the mid-term exam, I was among the very few to succeed in solving the problem and getting a good grade. However, even though the final exam was of exactly the same type, I failed miserably at it.

The last teacher I shall mention here was Paul Schleyer (1930–2014), whose graduate course in organic reaction mechanisms I studiously audited at Princeton in 1962–63. What was outstanding about it was the amazing extent of the attendant bibliography. Paul had read everything pertaining to the subject matter, which he presented both clearly and forcefully.

If I became a scientist and a chemist, a teacher as well, I owe it to the remarkable men and women described in this chapter. What, in my opinion, then contributes to the make-up of a great educator? An unconventional mindset, capable of maintaining the attention of students. The uncanny ability to self-project. A voice that somehow connects inside the mind of students. A charismatic personality, that induces admiration—creating the desire in students to raise themselves to the expectations of the teacher, however high.

Premises from Childhood and Adolescence

3

Georges Bernanos (1888–1948) wrote "*Si je recommençais ma vie je tâcherais de faire mes rêves encore plus grands; parce que la vie est infiniment plus belle et plus grande que je n'avais cru, même en rêve.*" (If I had to start my life over I would strive to make my dreams yet bigger; because life is infinitely more beautiful and greater than I have ever believed, even in dreaming).

My whole being agrees. Moreover, Bernanos forcefully insisted that we need to connect or to reconnect with *l'esprit d'enfance*—the childhood mind or spirit, a French expression whose translation does not convey the full meaning. This chapter details how my life as a scientist has its roots in childhood and adolescence.

I stood apart from many other children in one respect, from Denise in particular: she and I lived in Grenoble across the street from one another and were best friends from age 4 or 5 and still are. She loved vacations and I loved school, biding my time during vacations, when learning was suspended, yearning that I was for school to resume.

Not that I did not adore play and being playful, another trait I owe to my father. Even though he was often introspective and sad, he had a superb sense of the feast, the festival, the joyful festivity.

Denise and I played together a lot. I recall games of hide and seek, hopscotch, showing off in costumes, riddles, newspaper and magazine competitions (we won one together), etc. With other children, we would engage in cops and robbers, in its numerous variants—of course opting for the outlaws side (Fig. 3.1).

My experience of chess is minimal and I suspect I would have been poor at it. French children play *jeu de l'oie,* a favorite of mine: a dice throw determines one's position on a circuit dotted with traps. The first player to complete it wins the game. My initiation into golf, years later, took place of all courses in St. Andrews, Scotland. The analogy to *jeu de l'oie* became there obvious to me. Both are metaphorical for life, with its traps—such as nowadays electronic gadgets, social networks, addictions—and the attendant disappointments.

The quest for the advancement of knowledge is no laughing matter but an admirable endeavor. Science is also play—a point I have expanded on in the

P. Laszlo, *A Life and Career in Chemistry*,
https://doi.org/10.1007/978-3-030-82393-1_3

Fig. 3.1 Denise Jacquemin, my lifelong and oldest friend, since early childhood,. A daring person as this picture, from a climb in the Vercors range, eloquently shows. Now retired, she was a professor at the Université Stendhal, Grenoble, in linguistics (permissions in process)

American Scientist monthly. Words are revealing. In French, "to play" translates as *s'amuser,* the converse of *muser*, with the meaning of "being idle." The *muser* verb is a cognate of *museau*, "muzzle" in English; it referred originally to a buck sniffing for a doe in heat. From the contrast between work and play, earning one's living or indulging in sex, in French it acquired the less graphic meaning of idleness. Science is thus a hybrid activity, being both work and play.

But back to my childhood. At first, reading children books, I was fond of funny characters, such as General Dourakine in the Comtesse de Ségur novel of that name, or the grumpy and gouty grand-dad in *Little lord Fauntleroy*. I loved *Savant Cosinus*, *Sapeur Camember* and *Famille Fenouillard*, three illustrated books— they pioneered comics in the 1890s—authored under the pseudonym Christophe by a leading French botanist, Georges Colomb (1856–1945). They were great fun.

But I came especially under the spell of Jules Verne (1828–1905) and his *Voyages Extraordinaires* series of novels. They fleshed out the maths and sciences from school. For instance, what Monsieur Francès taught about time zones and the dateline came alive in *Around the World in 80 Days*; his explanation of longitude and latitude, of the skills of navigation using a sextant and a clock, were embodied in *The Children of Captain Grant*, a story of children looking for their stranded father around the globe, along a whole parallel. I admired geographical exploration and explorers: Roald Amundsen (1872–1928), Frijtdorf Nansen (1831–1930), Jean-Baptiste Charcot (1867–1936) and his doomed "Pourquoi-Pas?" boat.

Physics-wise, Jules Verne was a precursor: the concept of black holes occurs in the above-mentioned *Children of Captain Grant*; their worldwide quest can be argued to occur in a time-space continuum, anticipating upon Einstein's relativity, which Laurent Schwartz (1915–2002) admirably explained in his "Mathematical models in physics" classes. Chemistry-wise, Jules Verne was very much taken with nitrocellulose as a burning powder and explosive, also with its use as collodion for bandaging wounds, etc.

My joint vacations with Denise took place in a village, Colombe, about a one-hour drive from Grenoble. A cousin of her mother, Madeleine Thuillier (1908–1981), owned a farm there. We would typically spend summer months on this farm. The house was a dilapidated former castle-mansion, complete with the remnants of a watchtower. It was set on a hillside in the so-called *Terres Froides* area, in-between Grenoble and Lyon. I cherish my recollections, not only of games with Denise, also harvests in the scorching summer sun, the harvester coming to the farm to process wheat, the huge ensuing meal for all the gathered hands, grape harvesting as the summer waned, our following the adventures of a detective in a Catholic weekly magazine … (Fig. 3.2)

Multi-varied life on the farm had remained stable for centuries. I term myself very lucky to have experienced it, with its long-gone features, such as male and female tramps roaming the countryside and being welcome guests at our lunch table; attending the huge centuries-old nearby Beaucroissant fair; Madeleine making her own butter, her own cheeses; impressive delirium tremens crises of Rémy Thuillier (1893–1949) her alcoholic brother; her devotion at the local Catholic church; impassioned Easter time sermons by a Franciscan monk; Denise and I looking after the grazing cows and milking them; hunting a cow who had escaped into the woods to give birth; witnessing the bull on the farm impregnating cows brought to him by other farmers from Colombe or nearby villages; goats and the hostile geese on the property; and so on and so on.

Another experience would prove to be seminal to my activities as a chemist, much later. There was, a stone throw from the farm, a small clay pit in a sunken lane. Denise and I would gather some clay, marvel at its plasticity, mold it into small artefacts—typically earthen pots and dishes—and sun-dry them. The *Terres Froides*—the Cold Dictrict—owed their name to the soil being rich in clay. Clay swelling makes it retain considerable amounts of rainwater, slow to evaporate given the impermeability of the soil, that accordingly remains colder and for a longer time than in neighboring areas. Hence the harsher winters characteristic of *Terres Froides*.

I shall return in this chapter to this acquaintance with clays. I'll turn now to another important step in my becoming a scientist. This was my acquisition of English.

During the same period of the late 1940s as my stays in Colombe together with Denise, I started taking vacations, summer ones typically, in England—to be more precise, in Southwell, a little city in Nottinghamshire. I was hosted there by the

Bramwells. My parents had signed me up for an exchange program with an English family. The Bramwells consisted of Nellie the wife, Austin the husband and Katherine the daughter. Katherine was older than I by a year or two. She came only once to Grenoble, as she had no interest in learning French. But her parents took a liking to me and I to them. They were my English foster parents, I was calling them indeed Mummy and Daddy.

Austin ran a shoe store on the main street of Southwell, at 30 King Street, with the house in the back. He also made shoe repairs in a small attached shed. Nellie assisted him in the shop. They were terrific hosts, very generous and loving, intent upon helping me discover England and showing me around: the sights in London, the Minster cathedral in Southwell, the steel works in Sheffield, Shakespeareana in Stratford-upon-Avon, the Lincoln cathedral, the seashore at Skegness, etc.

England at that time still utterly differed from France. The food was totally new to me. The English were still under post-war rationing. Daily life had a different content, in the type of meals, their schedule, work and play activities. The townscape itself was made of sights novel to me, whether fish- and-chips stores, newsagents, the huge red mailboxes or yet a bowling green. The social structure enforced strict separation between the middle class (to which the Bramwells belonged) and either the lower middle class or the upper class. Needless to say, about to see Elizabeth's crowning, I became fascinated with the British monarchy.

Yet another difference struck me, the British tolerance and even fondness for eccentricity. French society by comparison was much more uniform and stifling. The Bramwells were such an example: Nellie and Austin were archers. I accompanied them to competitions. Their bows were much too stiff for me to handle though, one had to be very strong to use them. Robin Hood was their patron saint. I was taken, in nearby Sherwood Forest, to the so-called Robin Hood oak-tree.

To them, darts complemented archery: every evening, we repaired to the shoe store, hung a dartboard and had a go at it. Most of the time, Austin, with a good eye and a very accurate throw, won.

I was something of a curiosity in Southwell as the only foreigner. Circumstances for my learning the English language could not have been better, it was total immersion. I opened up totally to that extremely different culture. The language was challenging, the spelling of a word does not imply its pronunciation. But I was fortunate with a good memory to draw on. During my second stay with the Bramwells, the following year, I suddenly realized one day I had started to both think and dream in English.

This was a precious acquisition: I entered the scientific community in the Sixties, when English was the lingua franca and leaders in my field of chemistry were all Anglophones.

The Bramwells did not speak any foreign language. They were representative in that of the British.

I discovered television at the Bramwells, which helped my English. The BBC still enjoyed a monopoly. It showed the fascinating to me game of Twenty Questions: a panel had to identify some object from so many questions. Some targets were exotic and recondite. The example I recall—I was flabbergasted they managed to identify it—was "a blotch on the escutcheon!" Twenty Questions introduced me to what being a scientist entailed: a wide-ranging curiosity, finding the most revealing questions to be asking, building in one's mind the most appropriate evolving picture from the answers. And of course being playful and creative.

Other key experiences prepared me, between ages of 8 and 18, to becoming a scientist. One was reading the monthly magazine *La Houille Blanche* in my father's line of work. I was fond of it because of a single feature by a so-called Professeur Cyprien Leborgne—an anagram to Neyrpic Grenoble. The pseudonym belonged to Pierre Danel (1902–1966). A graduate from *École Centrale*, he headed the small group of engineers at Neyrpic, to which my father belonged (in spite of being paid as a draughtsman). My father termed Danel a genius. He was a scientist amongst the engineers, he could have made a career as a successful academic.

Why did I like so much his chronicles in *La Houille Blanche*? Entitled "*Miscellanées*," they belonged to *science amusante*, they were whimsical and witty—to me a joy. Examples were:

1. A marble sarcophagus, resting on the ground on two small blocks, with a lid itself in marble, is the seat of a curious phenomenon: water gathers there, practically pure and

similar to distilled water. It provides on average three hundred liters per year. How can this be explained?

2. Generally speaking, a diving bell, in like manner to a ludion, cannot be stable at any level in water. In which case can it attain a stable equilibrium?

3. Divers had been at work for some time, to recover the cargo from a wreckage, when one of them came to the surface showing signs of altogether exceptional exhaustion. A nap quickly put him back on his legs and he was able to resume work the next day. Once again, he showed the same acute fatigue which went away after a night of rest. The next morning he went back down as usual and this time found himself absolutely incapable of standing upright after the ascent. He collapsed on the dock. His fellow-divers surrounded him, very worried by his condition. Unscrewing his helmet in a hurry, one of the rescuers sniffed and looked extremely puzzled. A familiar scent had reached his nostrils. He sniffed once again. The others watched in amazement. " What's wrong?" they asked. "Whisky!" whispered the kneeling man, convinced that his sense of smell had betrayed him. They all sniffed. The smell was undeniable. "He's drunk," said the first. The notion lacked common sense. "But how? . . ." asked another. This was the question that thwarted their understanding. How was it possible for a diver to get drunk underwater? The mystery would have delighted Sherlock Holmes. True, there were crates of whisky in the wrecked freighter, but the diver would have drowned while trying to drink it. He was imprisoned by his equipment. So what could have occurred? They did not say a word to the sleeping diver but when he went down the next day, another diver discreetly followed him.

What did he find out? Knowledgeable readers, you can guess the answer, if not, wait for it in a future issue.

I loved such little puzzles and their solutions. They belonged or referred back to the fields of hydrostatics and hydrodynamics, those in which readership of *La Houille Blanche* was knowledgeable. The attraction to me, I now realize, was the notion of science being fun, great fun! The world around us is replete with puzzles. Typically, an intriguing problem of the everyday, calling for its solution upon the properties of water and some engineering know-how. The scientific mind—that of the so called Cyprien Leborgne—rejoices in both raising and solving such problems. About that time, late 1940s and early 1950s, I discovered similar puzzles and their solutions in equally enjoyable little books in English by George Gamow (1904–1968), the "Mr. Tomkins" series. I was lucky to read the small pieces by Cyprien Leborgne and George Gamow as a little boy entering my teen years, they deeply influenced me.

At this point of my recollections, some thoughts about the benefits of learning a foreign language. Hearing my parents speak Hungarian at a very young age while not understanding a word of their dialog was formative. It somehow wired my brain to seek meaning in the strange sounds they uttered. Hence, even though I do not speak that admittedly difficult language, I am able to pronounce the *gy* sound in *Magyar* or *nagy*.

Several stays at the Bramwells gave me English as my second language. As a weak bilingual although not a true bilingual, my French accent in spoken English betrays me.

The experience however was formative. When the Laszlos moved to Brazil—I was 12—learning Portuguese was easy: as the written rather than the spoken language, though: total immersion, in that case, came from reading novels—Jorge Amado's (1912–2001) in particular. Limited, perhaps 15–20% understanding, at the start, gradually improving and reaching 95% by the end of these first Brazilian books.

Having thus experienced the virtue of being bilingual, I have passed it on to my two eldest children, both trilingual—and much better at languages both written and spoken than I am.

My mind spontaneously seeks meaning in written signs. Alphabets are an example. Having skipped the first grade, I was not formally taught the French alphabet. It was no trouble whatsoever to learn the letters. But it took me several years, until I was about 10, to memorize from repetition their sequence.

I did not study Greek at school. But the Greek alphabet became familiar through its use in maths. Since I mention maths, I liked algebra best because of the self-same feature, seeking meanings in signs and symbols. When much later I first visited Russia (1977) for just a few weeks, the Cyrillic alphabet was a joy to encounter and learn to decipher. Later on, I discovered likewise the Kanji letters in Japanese (1979). Whenever, I visit that country, 15 or 20 Kanjis quickly come back to mind. A more exotic case is Tamahaq, the language of the Tuaregs: my notebook in 1981 when Martine and I joined a camel-mounted trip to the Sahara, through the Hoggar mountains, shows me jotting down some of the letters from that alphabet.

Interestingly, my childhood friend Denise (b 1937) had an intellectual journey similar to mine (her Dad was Polish): she entered linguistics and became a professor. Since I mention that field, I was always interested in it. To the point of serving a visiting professorship at the Department of Linguistics of the University of Chicago. My host there, John Goldsmith (b 1951), remains a very good friend.

What has all of this segment about learning languages have to do about my future career as a scientist and as a chemist? I taught myself the language of chemistry, that of structural formulas, as yet another foreign language. It is a language of its own. One might term it the Esperanto of chemists. When two chemists meet and they have no language in common whatsoever, they can nevertheless communicate, with pen and paper, writing structural formulas.

I'll end this chapter by mentioning yet another summer vacation, that awakened something of a passion in me, besides being delightful. I was in my late teens by then. Martine, my future wife, and I served a one-month internship in the Ratilly castle, in the Yonne (North Central France). It was owned by the stoneware potters Jeanne Pierlot (1917–1988) and Norbert Pierlot (1919–1979). Jeanne—from the Boutet de Montvel family of artists—had learned the craft from serving an apprenticeship with a professional potter, Eugène Lion (1867–1945), in Saint Amand en Puisaye, a French center of stoneware pottery.

Her husband Norbert, born and raised in Algeria, had been an actor and retained a close friendship with the mime Marcel Marceau. He switched to pottery at his wife's urging. Not only did he love the craft and excelled at it, he became a connoisseur of stoneware pottery from the Far East, Korea and Japan. He loved telling us about *yunomi,* the Japanese tea cups. A gifted teacher, he loved sharing his knowledge with us, as his students.

He had been deeply influenced by Bernard Leach (1887–1979), the British potter who had lived in Japan and absorbed there traditional Japanese pottery skills, which he wrote about—since the 1950s when I spent that month in Ratilly, the Leach story has been debunked by Edmund de Waal. A Japanese contemporary potter admired by Pierlot was Hamada Shoji (1894–1978).

But I better say a little more about my initiation into pottery. Hence, back to clay—an English word cognate of French *terre glaise.* In like manner to flour admixed with water turning into dough, dried clay is a white powder whose hydration turns it into putty. Which connected back with my delight in Mademoiselle Guillet's classroom playing with *pâte à modeler* (plasticine); or later in Colombe, together with Denise, from molding clay: material with the amazing ability, plasticity, to assume and retain whatever shape you give it.

In the Pierlots' workshop, we were initiated into using the potter's wheel. First, one had to learn how to center a lump of clay on the revolving wheel. It trained the brain and the hand into relishing the moment when centrifugal forces vanish as the piece of clay becomes centered. Then and only then can one start giving it a shape. Whenever afterwards I had the opportunity to test a potter's wheel—a Japanese brand low on the ground one crouches near became popular in the 1970s, I recall using it in the Middletown CT studio of Mrs. Betty Tishler, wife to Max Tishler who chaired the Chemistry Department at Wesleyan, she was a potter; I experienced again there this exquisite sensation from the tiny achievement.

Another likewise eerie feeling is to discover how uniformly unoriginal the shapes are that spontaneously come out from one's hands. The brain must be replete with a plethora of stereotypes it has recorded unconsciously. These get resurrected from the clay held on the wheel. To be creative, at least in my experience, demands a conscious effort and deliberation to steer clear from such clichéd shapes.

Once you have turned out a piece, a jug say, it is then left to dry for at least a week or 10 days. After one brushes it with the ingredients of a glaze, it then awaits its turn to be fired. Norbert Pierlot was not only a gifted teacher, he had been an actor—which he displayed in telling us the kiln mysteries.

He would stand by the kiln overnight, continually feeding the fire with small pieces of wood, relishing his anxiety at the eagerly awaited results. Heating the pieces to 1200–1300 C, as the stoneware clay required, had unpredictable consequences, he maintained. A few pieces were ruined, in one way or another. But others would mysteriously, depending on their location into the kiln, become anointed with a glaze that would be a joy to the eye and to the touch.

I was already a student of chemistry at the time of the stay at the Ratilly Castle with the Pierlots and learning from them the rudiments of pottery. Norbert Pierlot kept telling me: you ought as a chemist to apply your knowledge to clays. Their properties, their plasticity, swelling, firing—all of that entails a great deal of chemistry. He was prophetic, it took me only 20 years or so to heed his advice, as will become clear further in this book.

Becoming a Chemist

4

With the exception of my mother's death, my life did not show abrupt turns or hairpins. It was rather smooth going throughout. Thus, embracing the career of a scientist and becoming a chemist were mere transitions, rather than mutations.

Important decisions in life are perhaps to be avoided. The twentieth century had enough collective calamities, the two World Wars primarily, to avoid adding to them dramatic events for an individual: this was at least my perspective.

Did I just drift into chemistry? One might say so; or, more accurately, see it as the result of an entire slow process over my teenage years and into young adulthood, which I shall strive to chronicle in this chapter. Thus an evolution. One that alloyed personal history with collective history: the Cold War and Sputnik, the Algerian War, the Vietnam War—mankind truly indulged in avoidable warfare during the twentieth-century—against that I experienced the Golden Sixties and the *Trente Glorieuses*, the Second Chemical Revolution, the Double Helix, structuralism, the Linguistic Turn, etc.

First and for your amusement, something of a digression. Reminiscing about my daydreaming. As a teenager, when old enough to try and imagine my future, I nurtured three recurring fantasies. I knew they had little chance of occurring, but I nevertheless entertained them. They were fictions, close enough to the real for their recurrence, but with enough of a dose of gratuity that I knew them to be escapist.

The first was for me to become a ranger, a *garde-forestier*. This would have fulfilled two loves of mine, of the outdoors and of watching and caring for natural species. I would have loved that profession for its complexity: so many parameters to be considered and mastered on a daily, even hourly basis: animal life resident in the trees, in addition to birds and squirrels many other species such as mice, lizards, ants, worms, etc.; parasitic plants, such as *Usnea* lichens, mosses, mistletoe, etc.; coexistence of the old and the new, i.e., plants and their necessary diversity; people coming by, from fishermen and hunters to campers and poachers. I have seldom had the opportunity to meet foresters but each such time they impressed me with their wealth of knowledge and abundance of wisdom. Great and good people!

The second fantasy was for me to become an *Éclaireur de haute-montagne*: this is (or was) an elite corps in the French Army. They scout the Alpine ranges, at high altitude. Their uniform had the appeal of simplicity, white skiing trousers and parkas, making them undistinguishable in fields of snow and ice. I imagined them patrolling the borders with Spain, Italy and Switzerland, being both tough and agile, in a no man's land above the marmots and jackdaws, at ease on skis or crampons, also with climbing gear, prompt to rush to the rescue of stranded mountain climbers, but also familiar with chasing down smugglers and hostile gangsters haunting the mountains.

The third, assuredly most prevalent of those fantasies, was for me to become a taxi driver. I knew it to be hard work, with long hours, unbearable tedium while waiting for a fare and a dose of danger. To me, the attractiveness of that profession was its very diversity, bringing one in close contact to so very many persons, able to talk to them and thus to find out a little about them and what made them tick.

Now to my opting instead for chemistry. "Boys will be boys:" I was no exception to the dictum. Was it more chatting with my school friends or being so much taken by the Jules Verne novels? I experimented a lot with these tools of the sleuth, at least in fiction, invisible ink and explosives.

Invisible ink: quite a few recipes were around. My favorite was use of lemon juice. To reveal the written message, to apply heat is enough, for instance by ironing the paper bearing a message.

Explosives and rockets. Jules Verne, in *De la Terre à la Lune*, a truly remarkable anticipation, Cape Canaveral included, uses four hundred thousand pounds of *fulmi-coton* (nitrocellulose) as the propellant. I never played with that substance. But I did handle some other explosives.

The first I'll mention is lycopod powder. It consists in the dry spores of a *Lycopodium* plant. When thrust into a flame, that of a Bunsen burner for instance, it creates a spectacular flash, with a great ball of fire and a detonation. I used this device for the last time in 2001 for a demonstration in the lecture hall Baker 200 at Cornell University, to the students in the Chemistry 106 course (science for non-science students) I taught there as a visiting professor. This hugely impressive explosive is also a favorite of magicians.

Why do young people love being scared out of their wits, in amusement parks for instance, where frightening attractions often feature? For the relief of having survived it, I submit. As an initiation rite. To show themselves and others how tough they are. Which is why this lycopod powder explosion became a favorite demonstration of mine.

Then, there is Berthollet's powder—whose recipe I owed to the already mentioned Professor Bahiana, at *Liceu Franco-Brasileiro*, in Rio de Janeiro. That powder, devised in the eighteenth century, is a simple mix of ordinary granulated sugar and potassium chlorate.

I started experimenting with it right there, in Rio, I was then 13 and 14. I used it also in Grenoble, after my parents and us, myself and my siblings, had returned to France from Brazil. I would purchase the oxidant, the potassium chlorate, from a local druggist. These were happy times when a merchant would sell such a dangerous chemical to a teenager, knowing full well to what use it would be put—not just for killing weeds in the garden.

The mixture burns beautifully. It lends itself to rather impressive explosive burns. I could not resist applying it to rocketry. This was when we lived in a house facing the beach at Urca, in Rio. It had a terrace overlooking the street that circled around the beach.

I had got hold of a hollow copper tube, about 20 cm (8 in) in length and 2 cm (0.8 in) in diameter. I filled it with Berthollet's powder and set it vertically in a makeshift launcher. When I lit with a match the piece of string I had attached to it, the ensuing events were quick. They were also terrifying.

The improvised rocket lifted itself slowly and almost silently. Then, instead of proceeding vertically, it swung downward and proceeded at top speed towards the beach, missing by a paper-thin distance the neck of a lady then walking on the sidewalk around the beach.

The event—which until now I have never told in print—scared me so much that it put an early end to my promising career as a rocket scientist. So much for my emulating Werner von Braun (1912–1977).

The next episode I'll recount illustrates the negative effects of idleness, at a rather young age. My luck once again: I term myself lucky not to have been summoned to a police station for my naughty actions.

I was 16. This was when I had to redo the terminal year of high school in order to enter *classe préparatoire*. The only classes I was attending were in maths and physics, I was excused from all the rest that I had already successfully mastered for the *baccalauréat*—the graduating exam at the end of high school. Hence, my weekly schedule totaled about 12 hours, instead of the normal 30 or so. Accordingly, I would stay in bed late every day, until about 10 am. And I actively sought trouble by roaming secondhand bookstores and other businesses.

This is how I purchased, at a photographer's, his remaining stock of magnesium powder. In yesteryear, journalists would be accompanied by a photographer. The electronic flash had not yet been devised. Accordingly, for an instantaneous black-and-white picture, the photographer would set fire to a small bag of magnesium powder.

Thus I became the owner of about 50 such pouches, each with its small cotton fuse. How did I use them? In the worst possible manner, typical of a naughty teenager. I enlisted the assistance of a close friend, Jean-Paul Pittion (b 1937)—he later became a *normalien*, a professor at Trinity College in Dublin and a world-renowned scholar in Renaissance books.

Jean-Paul and I would surreptitiously enter one of the numerous apartment buildings in Grenoble. We would walk up the staircase. Each story had two or three apartment entrances, each with its doormat. We would deposit a magnesium flare on a doormat and ring the bell—we were quick to learn how many seconds the answer would take and we would time accordingly the fuse. We did it at lunchtime, when the apartment was certain to be occupied. The victim would open his or her door only to be met by a big flash of light and explosion.

In addition, while Jean-Paul and I would set our little devices, one of us would unscrew the bakelite top to the bell switch and remove it. We thus acquired several dozen entrance bell covers. Jean-Paul stored them at home in a cupboard. When his Dad asked him, "what is it you keep in that box?", Jean-Paul had to confess. His father was not pleased.

Once, during these expeditions in apartment buildings, Jean-Paul and I noticed a small round window in the stairwell, it was partly opened and the light inside was on: obviously, an occupied toilet. We lit and threw in one of our magnesium flares. I can still hear the shrieks of the terrified lady!

Berthollet's powder and magnesium powder were not the only chemicals I became very familiar with as a teen. There was also nitrogen triiodide. It is readily made combining iodine and ammonia, which provides one with a wet mash. It detonates as soon as it is dry. This was great fun too. One of my recollections is that the owner of a small movie theater, in the center of Grenoble, next to Place Grenette, had raised the entrance price for students. After his theater was seeded with nitrogen triiodide for a day, he relented.

Clearly, I owe my experience with explosives to Senhor Bahiana, he lit (pun intended) an interest in the boy he selected as his assistant and demonstrator. It left me with both a genuine interest in the magic of chemicals and a measure of composure while handling dangerous stuff. No, I did not become an expert at mine-clearing, which I might have. But, in later years, as an adult with a family of my own, for many years I would celebrate the passing of the year with fireworks of my own, in the garden, to the delight of the children. Of course, this was a show for which I did not seek permission from the local authorities.

Boys will be boys, as I wrote earlier. The fascination with explosives extends to poisons, another danger posed by chemicals. Professor Deluchat would convey it wittily: he would say "sulfuric acid is very toxic, a drop in the eye of a dog will kill the sturdiest of men"—a silly apophtegm if there is one, but that his students relished, as a break in the otherwise overabundant subject matter. I did not, by the way, risk such a test!

In this account of my early brush with dangerous chemicals, I mentioned nitrogen tri-iodide. A nice name, don't you think? Probably not. Anyway, I like it. Without going into the reasons, chemical terminology also had its attractiveness for me—and was among the reasons for my becoming a chemist. Other examples of names I found to be seductive?

Mercurochrome: this was a bright red liquid that served in my childhood as an antiseptic, on small skin abrasions or cuts. Regarding red color, the word vermillion, i.e., mercury sulfide, I still find fascinating. I loved that chemistry was a very modern science, but one with a rich past going back to Antiquity. Ammonia for instance: the name refers to the Egyptian god Ammon. Or, among the tools of the chemist, matras, a small glass vial whose name derives from the Arabic language: as with other sciences, mathematics in particular, the Arabs were a determining force in the rise of Western science, especially during the heyday of Baghdad as the main center of culture between the eighth and tenth centuries. Another nice term, in French, is *bain-marie*, the word for a hot water bath. It refers to Mary the Egyptian, a mother figure to the protoscience of alchemy, that preceded chemistry. I love the French phrase *vert de gris,* verdigris in English. It denotes the color of oxidized copper. It is a misspelling of *vert de Grèce*, i.e., literally Greek green. It was originally prepared by hanging copper plates over hot vinegar in a sealed pot until the metal acquired a green crust.

I also owe to Senhor Bahiana my strong interest in the history of chemistry, he often referred in his classes to the Chemical Revolution of the eighteenth century, to Lavoisier and his contemporaries, such as Berthollet.

Chemistry was thus attractive to me because it was rich with history, with the legacy of long past times, that it thus preserved: *À la Recherche du Temps Perdu.* In addition or complementing this Proustian dimension, the time dimension of chemistry held its fascination to me. While its processes occur in wide-ranging times from femtoseconds (10^{-15} s) to years and centuries, lab work in chemistry offers gratifyingly quick answers. Whereas an experiment in physics may take months to set-up, you can get to your lab in the morning, devise an experiment that came to mind overnight, and receive an answer by the end of the day. I believe that some of the seduction of chemistry to me came from it being both a science and a craft.

Another dimension of chemistry that impressed itself on my mind during my teens was its relationship to the visual arts, to painting through pigments—such as cadmium yellow, cobalt blue and many others—and to photography, through silver salts. During my late teens, I visited regularly the Louvre Museum. Earlier on, I had acquired a preference for the Flemish so-called primitives, such as Memlinc, the Van Eycks, van der Weyden especially. As for photography, while still in Grenoble where we had set a small darkroom in the house, my second mother Ella and I spent long evenings enlarging black-and-white photographs. I loved the moment of truth, when the image rises on the paper, faint at first but quickly invigorating—a witnessing of chemistry in the doing. I was friendly also with the Gargat brothers, professional photographers (Fig. 4.1).

What kind of a person was I, say at age 18 when I started the NSE *classe préparatoire* at Lycée Saint-Louis in Paris? A politically aware person, with convictions of the Left, reading the daily *Le Monde*, the weeklies *L'Express* and *Le Nouvel Observateur,* when visiting England the *New Statesman*. A listener of classical music, on Sundays a faithful concertgoer attending at Salle Gaveau the baroque music concerts by Fernand Oubradous (1903–1986) and soloists of the ilk of Jean-Pierre Rampal (1922–2000), Pierre Pierlot (1921–2007), Marie-Claire Alain (1926–2013), etc. An avid reader. I had already under my belt, so to say, the contemporary American (Faulkner, Hemingway, Dos Passos, Steinbeck, Saroyan, ...) and French (Jules Romains, Martin du Gard, Saint-Exupéry, Malraux, Camus, Sartre, Vercors, Gary, ...) novelists. Borges was a great favorite of mine. Another major discovery to me was T.E. Lawrence's *Seven Pillars of Wisdom*. A bookish person, reading regularly the *Times Literary Supplement*, the *Lettres Nouvelles*, the *Quinzaine Littéraire*, ...

Philosophy was also a powerful interest. I had read already Kant and Spinoza. Also Sartre. Wittgenstein would come a few years later. At that age, in NSE, we had Jean-Toussaint Desanti (1914–2002) as our teacher of philo. He made us read La

Fig. 4.1 My second mother,
Ella Laszlo, in 1949
(photograph by the author)

Mettrie's *L'Homme-Machine* (1748). More to the point for this chapter, I became an
avid reader of Gaston Bachelard (1884–1962): his books were published by PUF,
their bookshop was across the street from Lycée Saint-Louis, and by José Corti, on
rue de Médicis across from the Luxembourg Gardens, a short walk from the Lycée. I
bought many of his books as soon as they came out. I found very engaging his
writings on chemistry, his enthusiastic endorsement of that science.

"All told, I shall be a chemist." Such a conclusion may have imprinted itself on my
mind, in my early twenties. Fortunately, the childish playing with explosives did not
leave me—or somebody else—maimed.

As for not following my fantasies with regard to a profession, allow me this brief
re-examination, from a rearview mirror so to say.

Forestry, with its constant multicentered scrutiny: chemistry would bring very
similar joys and quiet exhilaration. In my career, I treated it in like manner to an
ecological pursuit: practicing it as a generalist, rather than a specialist, truly lavished
on me great overall satisfaction.

Scouting on mountain tops? By operating at the leading edge of the discipline,
indeed I experienced a rarefied atmosphere. I was able to rely on comradeship of
other pioneers, and that was a constant joy.

What about driving a taxi? I have already spelled out what I anticipated from such a job, meeting a very diverse cast of characters. Lecturing about the advances in the lab on several continents, from Siberia to New Zealand, Canada to Australia, Brazil to Japan, allowed me to meet indeed all kinds of fellow-scientists. I am lucky for my professional life to have made me part of the scientific community, I can hardly think of a more gratifying human experience.

I have spelled out in this chapter the call of chemistry, as I felt it in my late teens before setting out to becoming an academic and a scientist in that field. There is yet another factor, one that I was not aware of for many years, until my early 30s. It was made clear, then, from a chance remark of my father: "you have become a chemist. Of all things, chemistry! This is the one subject I have never been good at." This is psychologically interesting, I had a need, mostly unconscious I believe, not to follow in my father's footsteps.

Learning the Ropes at Princeton 5

My wife Martine and I came back to Paris during the period 1963–1966 to allow her to complete her medical studies there. I served as an instructor at the new university in nearby Orsay, south of Paris. I have recounted elsewhere my work there in nuclear magnetic resonance—the then brand new tool that revolutionized chemistry.

French science in those years very much rejected Anglo-Saxon influences. A major factor was the widespread ignorance of English, reinforced by De Gaulle's edict to exclusively publish and communicate in French. Such monolingualism was rather crippling, assuredly a recipe for mediocrity. I was lucky to escape it, my postdoctoral year doing nmr at Princeton University in Paul Schleyer's group (1962–63) allowed me to be fluent in scientific English. It introduced me to the delights of living in Princeton, together with the very American moral imperative of "making a contribution." (Fig. 5.1)

My scientific life in Paris was spent within a small group of organic chemists striving to keep up with the rest of the world. What did this group consist of? The staff of the Institute of Natural Products (ICSN), headed by Edgar Lederer (1908–1988): he invited word-class lecturers for the weekly colloquium. Also from ICSN, the mechanistic team of Bianka Tchoubar (1910–1990) and Hugh Felkin (1922–2001) that Lederer brought in. In addition, the chemists at Collège de France where on Saturday mornings Alain Horeau (1909–1992) gave a superbly documented lecture. Also, the chemists-pharmacologists from the Roussel-Uclaf company. Within that group, almost clandestinely we read the textbooks by the Fiesers from Harvard. Summers brought the opportunity to join with other modern-minded colleagues, from Grenoble and Strasbourg especially, at the GECO meetings initiated by Guy Ourisson (1926–2006), on the model of Gordon Conferences in the US. I also attended the quantum chemistry conferences in Menton organized by Raymond Daudel (1920–2006) and Bernard Pullman (1919–1996).

To his puzzlement, I turned down Edgar Lederer, my supervisor, when he offered me a research position at CNRS. I was intent upon both teaching and research. Regarding the former though, I renounced preparing the *Agrégation* licensing exam.

P. Laszlo, *A Life and Career in Chemistry*,
https://doi.org/10.1007/978-3-030-82393-1_5

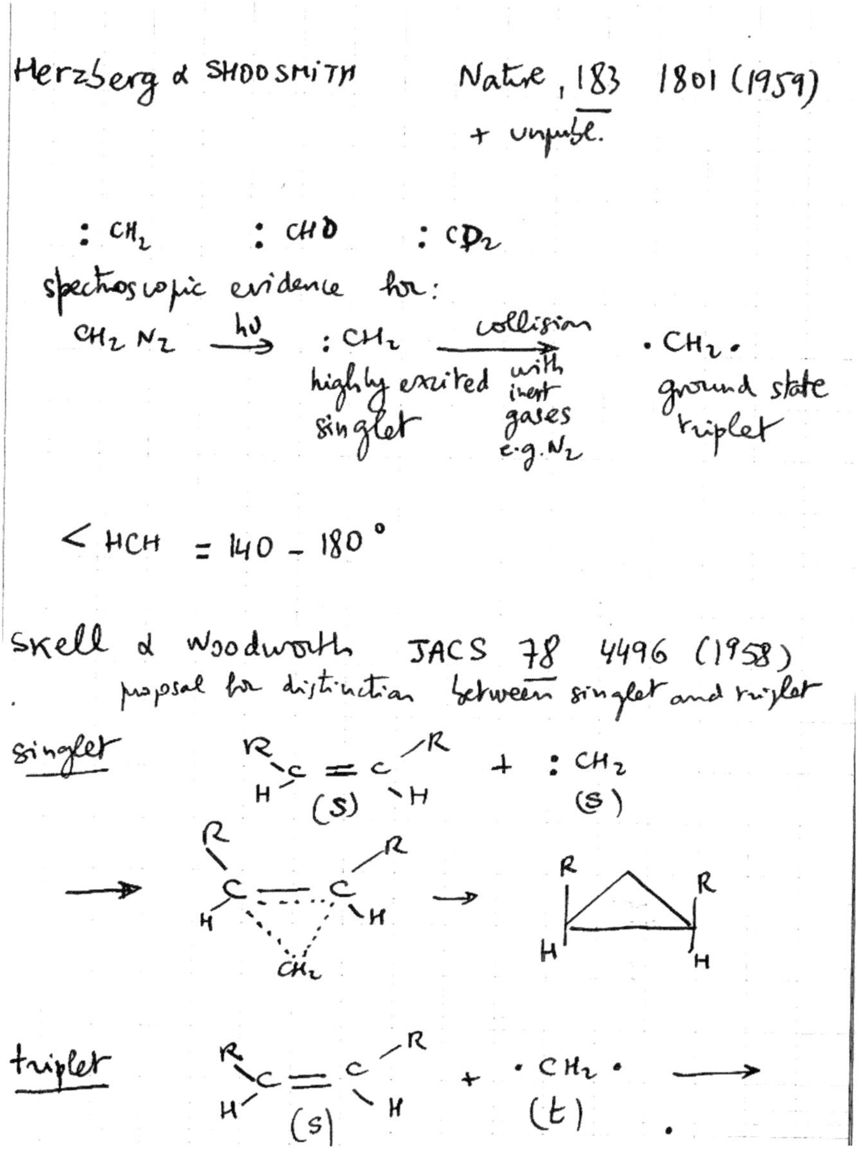

Herzberg & SHOOSMITH   Nature, 183 1801 (1959)
                                    + unpubl.

: CH_2 : CHD : CD_2
spectroscopic evidence for:

$CH_2 N_2 \xrightarrow{h\nu}$: CH_2 $\xrightarrow{\text{collision with inert gases e.g. } N_2}$. CH_2 .
              highly excited   ground state
              singlet   triplet

$\angle HCH = 140 - 180°$

Skell & Woodworth   JACS 78 4496 (1958)
        proposal for distinction between singlet and triplet

singlet

$$\underset{H}{\overset{R}{\diagdown}}C = C\underset{H}{\overset{R}{\diagup}} \quad (s) \quad + \quad : CH_2 \quad (s)$$

$\longrightarrow$

triplet

$$\underset{H}{\overset{R}{\diagdown}}C = C\underset{H}{\overset{R}{\diagup}} \quad (s) \quad + \quad . CH_2 . \quad (t) \quad \longrightarrow$$

Fig. 5.1 My notes from reading an article on methylene, during the 1962-63 winter

One evening, I came home tired out and hurting all over. I explained to Martine that a truck filled with laboratory furniture had arrived at lunchtime and that I had had to do the unloading. "You have not received a high level education to waste it like that," she commented. "Let's return to America." Thus, I sent out letters of candidacy to MIT, Oregon and Wyoming; and asked for letters of recommendation from the Chemistry Department at Princeton. Instead of honoring that request, three weeks later, I received an invitation as an assistant professor at Princeton.

Thus I needed an immigrant visa. When receiving it at the American Embassy, the Consul, a Harvard man, warned me that I might be drafted to fight in Vietnam; he then teased me about joining Princeton. Luckily, the threat from the Vietnam War was mooted by a letter to the draft board in Trenton sent by Robert F. Goheen (1919–2008), the Princeton President, asserting I was essential to the University. Since I have just mentioned him, whether as scholar or as administrator, Goheen was a great man—arguably the best President Princeton University ever had.

When I joined the Chemistry Department, Walter B. Kauzmann (1916–2009) was its chairman. He had replaced Donald F. Hornig (1920–2013), who after serving as the scientific adviser to John F. Kennedy, became president of Brown University. I was not the only recruit, far from it: there were 14 of us, newly appointed assistant professors. As it turned out, only three became tenured as professors. Most of the others however, enjoyed successful careers elsewhere.

I will break occasionally the narrative of my formative years at Princeton with small vignettes, to convey some of the flavor of our years there (1966–1970).

> From our previous stay (1962–63), we knew that we were expected to serve French cuisine and wine to friends and colleagues during dinners in our home. Accordingly, before we left Paris I bought a whole trunkload of bottles of wine, both red and white: good vintages and worthy origins. After I had retrieved the trunk from Customs in New York and opened it at home, I found the contents had been ransacked. The best bottles were gone. Corrupt longshoremen or customs officers, in league with wine dealers, had removed about 10% of the contents. For instance, the Meursault whites were gone. Our plunderers however, not only highly knowledgeable about French wines, were also very considerate: they had carefully stuffed the holes of missing bottles with crumpled copies of the *New York Times* of that day.

At that time, the fall of 1966, the department had a policy devised by some activist full professors. They had their eye on the then number one department, that of Harvard. Their ambition was to become a carbon copy of Harvard. Accordingly, many of the newly hired junior faculty, were Harvard graduate students or postdocs.

The major problem of the chemistry department at Princeton, not from hindsight since I realized it already at the time, was its small size, below a critical number of faculty members. A couple of years later, the defection of the biochemists, led by Jacques R. Fresco (b. 1928) and Charles Gilvarg (1926–2013), would only worsen the problem. The "to be Harvard-like" goal ignored what had made the Princeton department outstanding in earlier years, excellence in physical chemistry, when Henry Eyring (1901–1981) and Hugh Stott Taylor (1890–1974) had led it. Overall excellence was an impossibility given its small size, less than 24 faculty members. At the time, Princeton University could boast number one status among American university departments in physics, mathematics and philosophy—but not chemistry, in which it ranked about number ten.

After Kurt Mislow (1923–2017) became chairman in 1968, he did his best to attract a major appointee to the department as a full professor. The greatest need was in inorganic chemistry, where we had Neil Bartlett, but he was about to be snatched up by Berkeley as a potential Nobelist. In my recollection, Mislow tried with Jack Halpern (1925–2018), Earl Muetterties (1927–1984) and Rowland Pettit (1927–1981), who all declined.

> The atmosphere in the Department was highly congenial and friendly. For instance, some of the junior faculty would occasionally—once a week maybe—play touch football at the end of the day. Once, as I was rushing to catch the ball, Bill Horrocks stopped me. And, to prevent me from escaping, he sat on me—all of his 200 or so pounds. Foolishly, I attempted to move—and tore a muscle in my back; one of the reasons for my lifelong bad back. Bill, now retired from Penn State after 40 years of teaching, lives on Cape Cod.

Some—perhaps even most—full professors in the Department were outstanding teachers: Walter Kauzmann and John Turkevich in physical chemistry, Hubert N. Alyea in general introductory chemistry, Kurt Mislow, Paul Schleyer and Ted Taylor in organic chemistry. Alyea specialized to such an extent in teaching that alumni remembered fondly his lectures and demonstrations. He was a sweet person, I fondly recall him whistling to himself in walkways of the Department.

Among the senior colleagues, John Turkevich (1907–1998) was one of the most interesting. A warm personality. A dynamo. A star in catalysis. Discreetly recruiting for the CIA. And, although he never mentioned it, a bishop in the Russian Orthodox Church.

In passing, religion was never mentioned by any faculty member in the Department, the implicit understanding was to keep science and religion totally separate.

> Alumni were a force because of annual contributions that gave the university a hefty endowment. Every year, at the beginning of summer, a weekend was set for their class reunions—in part drinking bouts under tents on the campus. Each class had a uniform featuring the university colors, orange and black. The main event in a reunion was the so-called P-rade: alumni would go through town in rank, the eldest leading, the most recent graduates at the end. It was an eerie sight to watch the most powerful men in America (no women yet) marching proudly together. Some classes would even tow a trailer with a caged tiger, the mascot animal. I vividly recall seeing a couple at the airport in Paris about to return to the US, both flying the Princeton colors!

My senior colleagues in organic chemistry—I was labeled a physical organic chemist—were Dick Hill, Kurt Mislow, Paul Schleyer and Ted Taylor. Richard K. Hill, a natural products chemist, was quite a gentleman, soft spoken and benevolent. After 15 years at Princeton, he found the pressure suffocating and moved to the University of Georgia, in Athens. The pressure on him was from the two stars in the Department, Kurt Mislow (1923–2017) and Paul Schleyer (1930–2014).

Kurt was a deep and original thinker with important contributions to stereochemistry. He loved to correct mistakes by others, whether in the literature—or in the midst of a lecture. A splendid speaker, he was also very stimulating face to face.

Paul had been my supervisor during my postdoctoral stay at Princeton in 1962–63. He spearheaded the invitation for me to return as an assistant professor in 1966. He saw the need to distance himself from physical organic chemistry when, in the late Sixties, this subdiscipline became eclipsed by synthetic organic chemistry. With his forcefulness, he pioneered the budding field of computational chemistry.

The computer was very much born in Princeton. Since books exist on that point, I'll only mention Veblen, Church, Von Neumann, Turing and Tukey all present at birth, so to say. In the late 1960s, the university had an IBM 7094 computer, bought for about three million dollars—the equivalent of about $20 million nowadays. The computer center was housed in the School of Engineering. One had to apply for access. Such mainframe computers became obsolete by the 1980s and 1990s, with the advent of personal computers.

Schleyer thus needed a hefty portion of computer time at Princeton, a demand that the Department and the University would not meet. Paul was power-hungry. He had his eye on a Nobel prize and sensed that computational chemistry might bring him one. Which explains his forcefulness, that rankled his senior colleagues, bent upon maintaining the clubby atmosphere of a group of equals. Hence, Paul to his chagrin had to leave Princeton for Erlangen, in Germany.

Paul never forsook his Nobel dreams. Realizing that he might not get one alone, he went for a joint prize, cultivating in succession collaborations with H. C. Brown (1912–2004), George A. Olah (1927–2017) and John A. Pople (1925–2004)—each of which would win a Nobel.

I had been Paul's protégé. When his days at Princeton became numbered, also when physical organic chemistry was no longer in the fore, synthetic organic having taken its place, my contract was not renewed. Which reflects a general reality of academic life, not devoid of its power plays and struggles: those explain at least in part, together with protection, endemic in academia, the high mobility of professors between campuses.

Ted Taylor (1923–2017), who was also a superb lecturer, had a considerably more discreet presence than Kurt and Paul. His field was synthetic organic chemistry. He succeeded Kurt as Department Chairman in 1973—to Paul's disappointment: Ted was not sympathetic to Paul's clamoring for prime access to the university computers and to his seeking prima donna status. Ted had great luck in his research: an heterocyclic chemist, his work on the obscure topic of pteridines—molecules involved in the colors on butterflies—led him to the discovery of the anti-tumoral Alimta. Ted donated to Princeton a major portion of his royalties, thus enabling construction of a brand-new building for chemistry.

What did I learn from being a faculty member at Princeton?

Everything, is the short answer. To elaborate a little, I learned how to best teach a class: make the effort to go to the primary sources in order to get the facts—build thus a comprehensive documentation—proceed to construct the lecture jointly with your audience, fully involving the students. This, I learned not only by doing it in front of very alert and critical people, also from by example from attending numerous guest lectures—at least four each week. Emulation played its role, I was awed by the admirably eloquent speakers at the monthly meetings of the full university faculty.

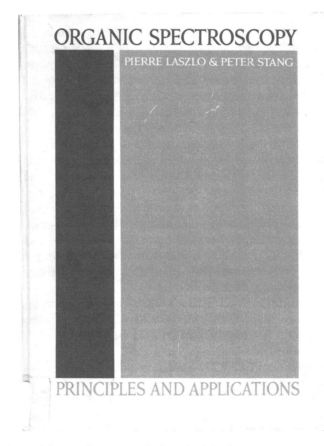

Fig. 5.2 Cover of *Organic Spectroscopy*, in the author's books

The main course I taught was organic spectroscopy, heavy on nmr. It was well attended, 20–25 people, mostly graduate students. There was in the Department a very smart graduate student, mischievous and histrionic, who rejoiced showing up in classes taught by various professors, disrupting them with falsely naive questions and sarcastic comments. He showed up in mine, I was able to hold him in check and he did not come back.

I also taught a graduate course on organic reaction mechanisms, that focused on tools such as isotope effects and acidity functions, in addition to kinetic studies (Fig. 5.2).

About 1970, Peter Stang (b. 1941), then a postdoc in Schleyer's group, now the long-standing editor of JACS, one of the two leading journals in chemistry, a member also of the National Academy of Sciences in the US, approached me suggesting publication of my lecture material in book form. I agreed, with the proviso that he chip in. Which he did. Organic Spectroscopy, our joint book and my first, was issued in 1971 by Harper and Row, in New York.

I also mastered the running of a research group in parallel to teaching. The best comparison is to a family, graduate students will do their best for you if you do your best for them. You must continually prove worthy of their respect. Thus, weekly group seminars are important, so that group members can prove themselves, in turn, worthy of your esteem. As an example, in my group at Princeton we read together F.A.L. Cotton's *Chemical applications of group theory,* a rather high-level small treatise first published in 1963.

Running a research group in experimental chemistry has its budgetary aspects. One example: we dried our solvents over molecular sieves—small stone pellets. They first had to be roasted in an oven. The commercial kind, as sold by chemical suppliers, cost at the time about a thousand dollars. There was a shop of supplies for artists, on Nassau Street, between Witherspoon Street and the Woolworth's store. There I found for less than $20 a small hot plate for ceramic jewelry, it did the job!

Another benefit came from my having been selected to run the Library Committee for the Department. It involved a joint monthly conference with Mrs. Johnson, the Chemistry Librarian, and a representative from the Central Library on campus. It taught me much about book publishing and how to select the worthiest titles, of lasting value and to leave out the more ordinary. I'll only mention here W. A. Benjamin, a New York publisher during (only) the 1960s, who benefited from the enlightened advice of the chemist J. D. Roberts (1918–2016) in putting together an amazing stable of great scientists as authors.

I served on another committee while at Princeton. This was not a university committee though, but one at ETS (Educational Testing Service), where tests and exams were produced for the whole American educational system, predominantly qualification for higher education. My brief was to help prepare a brand-new examination in the French language directed at graduate students; at Princeton in chemistry, for instance, they had to satisfy a requirement for both compulsory German and a second foreign language of their choice.

This ETS committee, half-a-dozen people strong, had a couple from ETS plus four or five academics. I recall Stirling Haig (b 1936), who went on to teach at the University of North Carolina, and Quentin Hope (1900–1982), who taught at the University of Indiana—the latter especially had an endearing familiarity with France and the French. He was the very first person who alerted me, at the time, to the excellence of Marc Fumaroli's (1932–2020) writings. Quentin had two sons. One would be part of a team searching for remains of killed American soldiers, world wide.The other, McKinnis Hope, served at the MacArthur Foundation, administering the so-called Genius Awards.

This collegial work for ETS taught me how to put together a multiple-choice exam, for which one needs to think hard in order to come-up with valid distracters. It would come in handy when my appointment switched from Princeton to Liège.

One learns by doing: my time at Princeton saw hard work, with 70-hours weeks and a great deal of organization to get things done and move along.

Life in Princeton

6

Princeton was a paradise in the late 1960s. An enclave within the US, the whole world—culturally-speaking—would visit Princeton. Thus, living there meant not only scholarly work under ideal conditions, also quality leisure; whether concerts, plays, lectures, readings (Fig. 6.1).

Some recollections are of the weekly paper, *Town Topics*, a great read, for instance because of a highly knowledgeable music critic, Arno Szafran if I remember his name correctly; of the concerts and plays at McCarter Theater; of boarding a bus to the Port Authority Bus Terminal in Manhattan and, upon return, glimpsing the tower of the Graduate College, a sight cherished already by F. Scott Fitzgerald (1896–1940).

This part of my account will include many names, some of whom were (or are) well-known. This is not to be mistaken for name-dropping. Not only are such names essential to my story, they also assist the process of rememoration, anchoring it so to speak.

I ran the weekly organic chemistry seminars held in room 309, on top of the Frick chemistry building. My job was to invite the speakers, I tried to select interesting people and topics; such as, for instance, Paul C. Lauterbur (1929–2007), who would devise MRI. These seminars were held evenings, the speaker was first plied with a fine dinner (and spirits). Were the latter meant to weaken him? Definitely yes: these seminars were modeled after those at UCLA, where the duumvirate of Donald J. Cram (1919–2001) and Saul Winstein (1912–1969) held a weekly show meant to humble the speakers—in the name, of course, of scientific truth. At Princeton, Kurt Mislow and Paul Schleyer sitting in the front row were the interrogators—the word is not too strong. One evening, a graduate student in the room loudly commented "will you leave the poor slob alone?"

These seminars imbued me, others also no doubt, with a weird talent. I refer to it as bifocal attentiveness, i.e., the ability to follow closely the speaker's argument while simultaneously working in your brain on a topic of your own: the presentation brings with it a special stimulation, it somehow enhances brain activity.

P. Laszlo, *A Life and Career in Chemistry*,
https://doi.org/10.1007/978-3-030-82393-1_6

Fig. 6.1 Together with my daughter Sophie, at Stanworth—the faculty housing we lived at in Princeton (winter 1966–7) (author's collection)

These seminars, which graduate students had to attend, served also as the source for some of the questions posed to them in the monthly cumulative exams (or cums). Graduate students had to pass a number of those (six?) during preparation of their Ph.D., i.e., in about three years. But these were tough exams. Graduate students feared them.

> During my time at Princeton, a graduate student indeed failed to satisfy the requirement. A fellow junior faculty member was adamant that he be expelled. But it was clearly a psychological block rather than a lack of competence or talent. I was eloquent and successful in my defense of this student, who went on to a highly successful academic career.

I need to repeat it: Princeton was an earthly paradise, for its unique mix of small-town America with cosmopolitanism: it hosted an academic community and a world-renowned university.

I rediscovered Firestone Library, of world class given the sheer size of its holdings, then about ten million books. The Faculty Club, later moved to the former residence of President Goheen, was on the top floor. After having had lunch at the nearby Student Center, I would join some colleagues there, usually from the nearby History Department.

> Martin Duberman (b. 1930), who was one of them, after I had asked how I could learn more about the US, made an unusually worthwhile suggestion, that I read Vladimir Nabokov's Lolita—one of whose aspects is as a road novel indeed.

I recall another time having much enjoyed a conversation with Oskar Morgenstern (1902–1977), the economist of lasting fame for having co-authored Game Theory with John von Neumann (1903–1957). He was urbane, highly knowledgeable and a kind gentleman.

We lived at Stanworth, a compound of small houses that the university had put up to house GI Bill students after World War II and had never been replaced with more durable structures. Among our neighbors, there were two junior faculty members from History: Michael S. Mahoney (1939–2008), an historian of computing, his wife Jean and their son Colin; Jerrold E. Seigel (b. 1936), an historian of intellectual life and culture, his wife the flutist Jayn Rosenfeld and their first daughter Micol.

Jerry and Jayn became lifelong friends, very dear to me. Jerry moved to NYU later on, but Jayn continued as a member of the small but distinguished Music Department at Princeton. Jerry and I would from time to time play tennis together on the university court, neither of us being a star player.

Another near-neighbor within the Stanworth Complex was the then Japanese postdoc working with Professor Frank H. Johnson (1908–1990) on bioluminescence. Named Shimomura Osamu (1928–2018), he and his wife Akemi had a son, Tsutomu, born in 1964 within a week of our oldest daughter, Sophie. She and Tsutomu were best friends, it was a joy to watch them racing their tricycles along the walks at Stanworth.

Dr. Shimomura Osamu won the 2008 Nobel Prize in Chemistry. He was born and raised in Nagasaki, where he suffered from the destruction wrought by the second atomic bomb to have been dropped on Japan in 1945. I vividly recall him confiding to me that he would never be able to forgive Americans this act of wanton cruelty.

With respect to world politics, yet another recollection from the years when we lived in Princeton and at Stanworth. Another near-neighbor was a former chief librarian from the Firestone. On Sunday September 4, 1966 he came over the lawn rushing to remonstrate with me, complaining bitterly about the speech De Gaulle had made a a few days earlier in Pnom-Penh, Cambodia, urging the US "as an old friend for many years" to get out of Vietnam. I tried, unsuccessfully, to placate him. The hostility to De Gaulle ran high in the US.

Friendships with fellow faculty members, many of whom were my age, were not the only ones I struck while at Princeton. I also made lifelong friends among students, in addition to the graduate students in my research group; and also among visitors of the campus.

Among undergraduates, who had chosen to prepare and write their senior thesis under my supervision, I'll single out Gregory L. Diskant (b 1948). At Princeton, he was on the staff of *The Daily Princetonian* from 1966 to 1970. From Princeton, having graduated in 1970, he went to law school at Columbia. He then clerked for Supreme Court Justice Thurgood Marshall (1908–1993) in 1975–76. He then became and was until 1980, an Assistant US Attorney in New York. He subsequently joined as a senior partner the law firm Patterson Belknap Webb & Tyler LLP, also in New York: he is a formidable trial lawyer, whom I was fortunate to have seen in action. Last but not least, he takes to heart his citizenship: noteworthy for an active pro bono practice focusing on democracy issues.

Another undergraduate from the Sixties, who has remained a good and admired friend, is David M. Pensak, who graduated from Princeton in 1969. He received his Ph. D., also in chemistry at Harvard with E. J. Corey (b 1928), the Nobel Prizewinner. After being the chief computational scientist at DuPont, until 2004, he founded Raptor as an Internet firewall in 2004; Revulytics in 2002 for development and marketing of a software anti-piracy product.

The third undergraduate I'll mention is John L. Soong, Jr. (b 1945) He graduated in 1967 and went on to medical school at Columbia in New York. He came to Princeton from Hong Kong, where his parents were Chinese grandees. John is a close relative of the three Soong sisters, who prior to the Communist takeover played important roles in twentieth century Chinese history. John became a radiologist and settled with his wife Suzie Marr in Honolulu, where she came from. About the end of last century, John wrote me a letter: his senior thesis had dealt with nuclear magnetic resonance (nmr). He was convinced he would never again touch that topic. However, when magnetic resonance imaging (mri) came to Hawaï, John was the only radiologist in Hawaï who understood how it worked; and he was called upon to often lecture about it to his fellow radiologists and physicians in the whole archipelago.

The undergraduates who chose to prepare their senior theses under my supervision were both outstanding and representative of Princeton at its best. The qualities of the Princeton men I came to know from close proximity were that of the American elite: very clever, articulate and attentive, good listeners, thoughtful gentlemen of few words and of keen judgment.

A delightful feature of life in Princeton was enactment on Sundays of a Bach cantata. All comers were welcome to rehearse one of the cantatas: most actively for both the instrument players and the singers. At the end of the day, the piece would be delivered. It was a splendid gathering to participate in. I believe it very much retrieved the original spirit of these wonderful compositions, the fervor of us amateurs somehow connected with the religious fervor in the eighteenth century.

I will turn now to some recollections from the year 1968. My luck had me invited to the University of Toulouse as a visiting professor during the month of May 1968, when French universities erupted in students protest. I gave a single lecture.

In the aftermath, I watched the events unfold. I spent numerous hours in lecture theaters, jointly with two biochemistry professors, Jean Asselineau (1921–2013) and Jean-Pierre Zalta (1924–2015). These were interminable collective brainstorming sessions, aiming for the most part at renovation of French higher education. I witnessed also a Napoleonic grab for power by a young chemist, unfortunately, not blocked in his move. Needless to say, all the previous power structure, the President of the university (the *Recteur*), the Dean of the Faculty of Sciences, had fled into hiding rather than assuming their duties. By the end of the month, the student collective in charge had run a telephone bill in the thousands of dollars, having made long calls all over the world! The country became paralyzed as workers went on strike as well. By the end of the month, having been barred from receiving any news from my group at Princeton, I decided to leave, jointly with my wife and daughter (who was then 4). We made the rounds of service stations and were able to fill the tank of our rental car and fled to Switzerland through deserted roads, only patrolled by a few *gendarmes*.

As soon as I got back to Princeton, I went to see Kurt Mislow, the then chair of the Department. He listened attentively to my story and made a few notes. A couple of days later, the Chemistry Department had reformed itself. Students representation was a done deal. What I had watched taking a month in Toulouse had been achieved, efficiently and without turmoil or long impassioned speeches, in just a few hours: American pragmatism and efficiency at its best.

Another set of recollections from the same year, a little earlier, at the beginning of April. I attended a national meeting of the American Chemical Society in San Francisco. This was the week when Martin Luther King (1929–1968) was shot in Memphis; the whole country was engulfed in riots, African Americans violently protesting his death. This was also the week when President Lyndon B. Johnson (1908–1973) announced he would not run for re-election.

In my hotel room, I was avidly listening to the news on television and reading the newspapers during that hectic time. Upon return to Princeton, I attended a service in the university chapel in memory of Dr. King. What most moved me was the singing of a psalm whose text had been written by Martin Luther.

In 1969, black students took over the New South administration building. They occupied it awhile. They thus protested the university's indirect support for apartheid in South Africa by its investments there.

I attended Commencement that year 1969. I was sitting close enough to President Goheen to watch him as he spoke. The events had weighed heavily on him and his hands were shaking. He was still young enough to fully recover though.

In 1973, thanks to his militancy, Princeton finally became coed. As for Goheen, President Jimmy Carter appointed him ambassador to India (1977–80).

But back to Princeton and the friendships I made there. Some of those were with visitors who had come to campus to recruit among our students.

There was Leon(idas) Petrakis (b 1935), who came to Princeton for the Gulf Oil Company, headquartered in Pittsburgh. Students protesting his presence—Gulf was synonymous with plundering the Third World and devastating the environment—besieged him in the Student Center. But Leon bravely addressed the protesters and walked through their midst. This was when our friendship started. Later on, when he chaired the Department of Applied Science at the Brookhaven National Laboratory (1989–1994), he invited me as a visiting scientist for a whole summer.

Another visitor at Princeton, also with a recruiting agenda, for DuPont in this case, was Fukunaga Tadamichi (b 1931), a Japanese chemist of the utmost talent, a former student of R. B. Woodward (1917–1979). At the Central Experimental Station in Wilmington, Delaware, he seconded Howard E. Simmons, Jr. (1929–1997) whom I have written about in a book about people who deserved a Nobel Prize in chemistry. Together, Simmons and Fukunaga did research work of the highest caliber, worthy of the very best academic institutions. We struck a friendship when he came to Princeton.

Even though he came to give a seminar, not to recruit, I'll mention next Sunney I. Chan (b 1936) from Caltech. I was impressed by how bright he was. His conversation shone with, for me, pearls of wisdom that came as advice. Later on, he arranged a visit for me to Pasadena. Much later yet, I saw him in Taïpei, when I was invited to lecture there at the Academia Sinica, where he had retired.

Of course, the name Princeton to many evokes not so much the University but the Institute of Advanced Study, that Einstein put on the map. I was fortunate to get to meet with a number of both residents and permanent members there.

Residents: in history, my lifelong friend Lester K. Little (b. 1935), historian of the Middle Ages, best known for his studies of monastic life, also of Justinian Plague, who met at the Institute his second wife Lella Gandini, an expert of children education. In mathematics, Hervé Jacquet (b. 1939). In theoretical physics too, Alex Grossmann (1930–2019) who shortly afterwards moved to the then new university of Marseille Luminy in France, where he was one of the devisers of *ondelette* (wavelet) theory, a powerful new tool in theoretical physics. He was the person who introduced me to the then brand-new *Dictionary of Scientific Biography*, whose devisor was another Princeton luminary, the historian Charles Coulston Gillispie (1918–2015).

Permanent members: Martine my wife (b 1938) was a resident (their first) at the small Princeton Hospital, unconnected to the University. When the art historian Erwin Panofsky (1892–1968) was taken with his final illness, Martine showed most generous and helpful compassion toward him and his second wife, Gerda Soergel (b 1929). We remained friends for quite a few years with Gerda, an art historian also, who taught at Drexel, in Philadelphia.

We also struck a friendship with the Weils, André Weil (1906–1998) the genius mathematician and brother of another genius, the philosopher Simone Weil, and his wife Évelyne (d. 1986). He was very talkative, often sarcastic about people or issues that he felt had deserved his dislike. One instance was the introduction of Social Studies at the Institute in 1973, that he strongly resented and fought hard to prevent. André Weil loved music—what can I say, he was transported by classical music. He would only listen to live performances, records were anathema to him. I vividly recall concerts at McCarter Theater, one would see his head bobbing in motion, as the music acted powerfully on him.

Another permanent member whom I recall was the Indian mathematician Harish-Chandra (1923–1983). During one of our meetings, he asked me about recent advances in chemistry. I proceeded to give him a picture of the entire field, as I saw it. He listened carefully, was obviously interested and asked pertinent questions.

To live in Princeton was to benefit from its attraction to non-residents: the whole world—not only academics, political leaders, performing artists, etc.—would come to visit and lecture, or perform. My vivid recollections include the writer Archibald McLeish (1892–1982) who read some of his poetry; the English philosopher Gilbert Ryle (1900–1978) for a series of lectures hostile to Descartes's views; the eminent linguist Roman Jakobson (1896–1982); the French actor and director Jean-Louis Barrault (1910–1994); the chemist Linus Pauling

(1901–1994); the future president of France, Valéry Giscard d'Estaing (1926–2020). Not to mention dinner parties where, among others, I met and had memorable exchanges with, among others, Ernst David Bergmann (1903–1975), chief-builder of the Israeli nuclear arsenal, or Hans-Magnus Enszensberger (b. 1929), the great German writer.

What was so special about living in Princeton? Being part of a community of scholars. It was exhilarating to be surrounded by those also committed to the advancement of knowledge. Usually, they shared similar political views to yours, were likewise interested in literature and the arts and in general avoided showing-off.

Have I painted too idyllic a picture? Certainly. It is imbued with my love for Princeton and the US.

Whenever we got together with fellow Europeans who also lived and worked in Princeton, the conversation converged to a single point, American were child-like: picky about their food—chatterboxes, easily vainglorious and boastful—as a rule, gregarious and monolingual—too money-minded—parochial—extremely naive and uncritical, for instance swallowing the lies of their government about Vietnam... And yet, totally endearing and adorable.

What I learned from my years at Princeton—admittedly too few—can be summarized with a dozen indices of the quality of university life: number one for me is the research group. That it consists of motivated and talented hard workers. Number 2 is for easy financial support of one's research. Number 3 is quality of the student body: bright, motivated young people. 4: the number of contact hours with students, it ought to be kept to 25–30% of one's time, preparation of classes included. 5 is seminars, that they be numerous and diverse, providing you with a weekly choice between at least half-a-dozen. 6 is the amount of the salary, that it be generous enough that you don't have to worry about your life support. 7 is the availability of a good library, if possible with an open stacks policy. 8 is the presence of friendly and stimulating colleagues (yes, I put friendliness before emulation). 9 is status, that the college or university ranks highly among institutions of higher education. 10 is attractiveness of the campus. 11 is the availability of state-of-the-art laboratories. 12, finally, the size of one's retirement pension.

Chemistry in the Sixties

The Sixties: these words evoke a period of turmoil, the Vietnam War, worldwide students protests whether at Berkeley or in France, drug addiction and assassinations of political activists. Was chemical science likewise torn apart? While it did not undergo revolutions, there were highly significant advances, along new directions.

One might see a symbolic portent of those in the shedding of the lab coat. It had been the uniform of the laboratory chemist who, henceforth, would wear jeans and T-shirts—putting on a white lab coat only to perform risky work at the bench or under the hood. The change in the attire reflected demographics, the impatience of the Baby Boomers with the rules inherited from previous generations.

The Sixties marked overwhelming American dominance of chemical science. This consequence of the Cold War can be illustrated by migration of English and English languages chemists to American universities: Frank A. L. Anet (b 1926), Neil Bartlett (1932–2008), Robert H. Crabtree (b 1948), Michael J. S. Dewar (1918–1997), Jeremy R. Knowles (1935–2008), Robert Langridge (b 1933), Rowland Pettit (1927–1981), John A. Pople (1925–2004), A. Ian Scott (1928–2007), Saul Winstein (1912–1969) were examples of this trend. They immigrated to the US not so much for the money, financial support of research, as for the stimulation from competition with colleagues who like themselves headed well-equipped laboratories and attracted hard-working and motivated graduate students.

What were some of the new directions imparted by both American-born and foreign-born chemists? Re-examination of the rules. One such was the so-called Octet Rule. Making novel noble-gas compounds, by the dozens, put an end to what turned out to be just a diktat from Linus Pauling (1901–1994) in his textbooks rather than a law of nature.

Terminology is important in science. Jeremy I. Musher (1936–1974), a brilliant young American chemist, who had been a Junior Fellow at Harvard, prior to joining the faculty at Yeshiva University in New York, coined the term hypervalency that put all such compounds under a single umbrella; and he gave the theoretical analysis to account for the bonding in such molecules.

P. Laszlo, *A Life and Career in Chemistry*,
https://doi.org/10.1007/978-3-030-82393-1_7

The Sixties were a time for chemical science to evolve under the impact of the computer, a relatively recent technological advance. The machines were mainframe monsters, universities purchased them from IBM typically and housed them in buildings, thus named computer centers.

I'll mention first some applications for an impact on chemistry, deep and durable. Chemists at Harvard University led in such applications of this exceptional new tool.

E. J. Corey (b 1928), from the chemistry department at Harvard, was already a star of organic synthesis in the Sixties, the devising and making of complex molecular architectures, with blueprints from natural products. He realized early on that the computer could help in the logical design of a synthetic approach to any targeted molecule. Since an actual synthesis works by binding together fragments, in like manner to a mason assembling bricks, Corey sought to list—in what he termed retrosynthetic analysis—all disconnections between atoms of the target molecule. These yielded a series of intermediate and somewhat less elaborate targets. These in turn were submitted to retrosynthetic analysis—and so on, and so forth, until one was left with elementary starting materials.

In Corey's Words (from his 1990 Nobel Lecture), "use of computers to generate possible pathways for chemical synthesis, which was first demonstrated in the 1960s, was made possible by the development of the retrosynthetic methods and the required computer methodology. Graphical input of structures by hand drawing using an electrostatic tablet and stylus, in the natural manner of a chemist, and output to a video terminal provided an extraordinarily simple and effective interface between chemist and machine." E.J. Corey, "the Logic of Chemical Synthesis: Multistep Synthesis of Complex Carbogenic Molecules (Nobel Lecture)", *Angewandte Chemie, International Edition in English*, 30(5), 455–465, 1991.

Of course, the computer can be used not only for storage of a large number of chemical structures, it can also be used for calculations, however elaborate. Thus, quantum chemistry underwent major advances from widespread availability of mainframe computers in the Sixties. A major advance was its use in so-called semi-empirical calculations.

Roald Hoffmann (b 1937) was a Junior Fellow at Harvard. The Society of Fellows is a superb institution for the encouragement of gifted young scholars. Roald—as a friend, I feel free to use his first name—collaborated with Bill Lipscomb, also a future Nobel prizewinner. He devised and first published in 1963 a semi-empirical quantum chemical tool, the Extended Hückel Theory (EHT). This computational method gave at the time invaluable access to molecular structures of all kinds. Why extended? The original Hückel method was applicable only to molecules with delocalized electrons along double bonds. Roald extended it to molecules in general, a big step forward. It was also the first semi-empirical theory to be made available to chemists. It certainly spurred other theorists, such as John A. Pople (1925–2004) and Michael J. S. Dewar (1918–1997) (both British) to devise other semi-empirical treatments slightly more refined—more about those now.

John A. Pople (1925–2004), a brilliant English theorist, thus devised CNDO and INDO approximations between 1964 and 1966, that enabled calculations of molecular properties for large numbers of molecules, hence became widely used by chemists after the attendant software was made available.

Another segment of this chapter deals with the Woodward-Hoffmann rules for electrocyclic reactions. Before turning to them, let me address the relationship of chemists to computers in the 1960s. Since this is an autobiography, I feel justified in taking my own example.

At the beginning of the Sixties and of my doctorate, working at the University of Orsay in the south Paris area, I studied the nmr spectra of natural products of the terpenes family. They presented a $CH=CH_2$ group, that showed a pattern of 15 lines, due to the interaction of the three hydrogen nuclei in the group—let us refer to them as A, B and C. As I was starting to analyze it, an issue of the *Journal of Chemical Physics* (*JCP*) arrived at the library. It featured a paper by Salvatore (Turi) Castellano and John S. Waugh (1929–2014) on the analysis of such ABC nmr interaction. This article answered my need, I set about to read it in detail and apply it to my problem.

Then—and this turned out to have been my only brush with programming—I translated Castellano and Waugh's equations into software, that I wrote in the then prevalent Fortran language. I punched both that software and the data into a set of cards that partly filled a shoebox. After several failures, due to deficient cards, my program ran successfully on the university computer, an IBM 650 model—a main-frame of course. This was an impressive bulky machine, although its performance is puny by today's standards. Thus, I obtained the parameters characteristic of such a $CH=CH_2$ group. (see Pierre Laszlo, "Wits and Smarts Make Scientific Pioneers," in Strom & Mainz, eds., *Pioneers of Magnetic Resonance*, ch 3, pp; 33-65, ACS Symposium Series, Washington DC, 2020).

In the Sixties, in my personal experience, the input for a computer was made of punched cards and the output was printed on wide sheets of paper, perforated on the edges for fast delivery.

I was typical also of my contemporaries, I believe, in switching to mini-computers when they became available: about 1977, I equipped my laboratory at the University of Liège with the very first Apple machines, six or eight of them and was able to start using a Macintosh as soon as it came out in 1984.

In the US and other countries, medical studies require a prior course in either biology or chemistry. The latter is centered on organic chemistry: pre-meds are thus given reams of facts for memorization, in the form of hefty textbooks with many hundreds of pages. The Sixties filled them with new developments that quickly migrated from laboratories to classrooms.

In addition, while the Sixties were a period challenging previous accomplishments—such as the already referred-to Octet Rule—novel chapters were added to the chemistry curriculum. This was the case of the Woodward-Hoffmann Rules, applicable to a class of transformations, known as pericyclic reactions.

These consist in a simultaneous—concerted—series of bond breaking and bond making events. Robert B. Woodward (1917–1979), a Harvard professor and a grand master in the synthesis of highly complex natural products, launched that of vitamin

B_{12}. Dorothy Crowfoot Hodgkin (1910–1994), the British X-ray crystallographer, published its structure in 1955. For this very ambitious task, Woodward enlisted the collaboration of the research group of Albert Eschenmoser (b 1925) at the Swiss Polytechnic (ETH) in Zürich. In so doing, Woodward became aware of pericyclic reactions and started cataloguing them. His intuition told him that their characteristic bond making/bond breaking, together with the specificity in product type, regarding spatial structure, had to have a theoretical explanation. He thus enlisted the help of a quantum chemist, the young Roald Hoffmann (b 1927). Together, they formulated the so-called Woodward-Hoffmann rules, a major expansion of the Hückel rules for aromaticity, i.e., special stabilization undergone by a class of molecules with alternating sets of ethylenic double bonds. These rules were a triumph of chemical theory—and of chemistry proper—during the Sixties.

The Sixties jettisoned Pauling's Octet Rule—only to "replace" it with these other rules, the Woodward-Hoffmann rules. As Woodward and Hoffmann set them in a book, they stated in its conclusion: "exceptions? There are none." A novel dogma to replace the discarded ancient one.

Pericyclic reactions were a major stream of activity and publications by organic chemists during the mid-1960s. Unfortunately, Woodward died too young, in 1979, to be awarded the second Nobel prize in chemistry he so clearly deserved. But Roald got his, in 1981.

There is widespread agreement that the era of organometallics was ushered in by the discovery of ferrocene and the visionary elucidation of its structure, for which Woodward was unjustly denied a share of the 1973 Nobel Prize awarded to Geoffrey Wilkinson (1921–1996). In 1968, Andrew Streitwieser (b 1927) and Ulrich Müller-Westerhoff 19xy-2019) would, by analogy, prepare uranocene.

The Sixties saw consequences of this seminal discovery. In 1965, M. F. Hawthorne (b 1928) made dicarbolyl iron(II), which brought attention to the boranes and carboranes. William N. Lipscomb (1919–2011) at Harvard vigorously pursued studies on their structures and dynamics, publishing the books *Boron Hydrides* (1963) and *Nuclear Magnetic Resonance Studies of Boron and Related Compounds* both books published (of course) with W.A. Benjamin (1969).

In 1964, Ernst Otto Fisher (1918–2007) in Munich made the first metal-carbene complex. Stabilization of a highly unstable and reactive species such as a carbene from its attachment to a metal atom was a fertile notion. Thus, for instance, cyclobutadiene became coordinated to iron (Criegee, 1959), which Rowland Pettit (1927–1981) exploited in his groundbreaking studies of iron carbonyls in 1965. His addiction to cigarettes killed Pettit when still relatively young. Was it in 1968? Pettit came to Princeton and presented a lecture series. He was brilliant and I was fascinated. Was it his dabbling into an extraordinarily wide variety of subjects, which was very much a temptation of mine too? In any case, I started work in my lab on metal carbonyls, a topic I later brought with me to the University of Liège.

As all those previous examples show, an organometallic complex consists, in the simplest case, of a metal atom coordinated by some ligands: a recipe trivially easy to enforce. Accordingly, the Sixties saw an orgy of organometallics—prepared for the sake of making a new compound. This rapidly became conformism, to which succumbed chemists in a least three advanced industrialized countries, Britain, Italy and France. The fad, of course, was mostly sterile. In my country, France, the ground had been prepared by science administrators who attempted to specialize universities into individual elements from the periodic table, such as sulfur at the University of Caen and phosphorus at the University of Toulouse—a rather narrow-minded intuition that of course did not bring forth important discoveries. The situation was different in England where Geoffrey Wilkinson taught at Imperial College and nurtured young talents, such as Malcolm L. H. Green (1936–2020).

One of Wilkinson's accomplishments (1967), together with John A. Osborn (1939–2000), was a catalyst that became named after him. In it, rhodium is the metal. It is coordinated to three triphenylphosphine and a chlorine ligand. It was historically the first homogeneous hydrogenation catalyst: since it is coordinately unsaturated, the metal center can bind an additional hydrogen molecule and/or an ethylenic substrate, hence its role and importance.

This species has only 16 electrons around the metal, which accounts for its coordinative unsaturation. Which brings up the 18-electron rule, the equivalent for organometallics of the Octet Rule in organic chemistry. The 18-electron rule turned into a guiding principle of inorganic chemistry, particularly in organometallic chemistry. As exemplified by Wilkinson's catalysts, four-coordinate square planar complexes do not follow the 18-electron rule and are found to be stable with only 16 electrons.

The massive emphasis on organometallics during the Sixties brought rewards, not only its lemming-like behavior on the part of too many: it helped chemists focus on the entire periodic table, on transition elements in particular. It made them aware of other types of bonding, in addition to covalent or hydrogen bonds: bridged, as in diborane; multicentric as in sandwich compounds such as ferrocene; and coordinative.

The Sixties saw the birth and rapid growth of bio-inorganic as a new subdiscipline of chemistry. I witnessed it from relatively close quarters. A major force propelling forward this new field was Professor R. J. P. Williams (1926–2015) from Oxford University. I came to know well Bob: not only did we have nmr as a joint favorite tool, we also had close common friends in both Raymond Dwek (b 1941) and Sture Forsén (b 1932). Bob had amazing energy. He was most articulate, in speech or in writing. He was a born organizer and he put together a most productive research group. Its publication output, in the form of both research articles and books was considerable—I resist using the adjective awesome, it has become trite from overuse. I am proud of having successfully nominated Bob for an honorary doctorate at the University of Liège.

The best way to summarize his accomplishments is that he explained in evolutionary terms how life forms continually dipped into the periodic table of the elements, at various times over billions of years. As oxygen came to dominate the composition of the atmosphere, they had to devise adaptations and to switch from some elements to others in their make-up. In the words of one of his disciples, "he gave chemical reasons why life uses phosphate to store information in DNA, why life collects magnesium to help stabilize that phosphate, why it must reject calcium to avoid solidifying that phosphate, and why it also must accept potassium and reject sodium to maintain osmotic balance."

The other pioneering bio-inorganic chemist I came close to was Joan S. Valentine (b 1945). She was a graduate student at Princeton when I was an assistant professor there. She moved to Los Angeles, as a professor at UCLA, where she has worked since 1980. She has worked on copper-zinc containing enzymes and their role in Lou Gehrig's disease (amyotrophic lateral sclerosis).

Bio-inorganic chemistry was a product of the Sixties, as part of an enthused invasion of the entire periodic table by chemistry. Nmr was one of the reasons, two-thirds of the nuclides in the periodic table being nmr-active. Another reason was simply the exponential growth in the number of chemists, to find an original slot for one's contribution the periodic table was an obvious move.

This helps to explain why the expansion of bio-inorganic chemistry has eclipsed that of bio-organic chemistry—the latter no less important, in the absolute.

During the Sixties, a contemporary and friend of mine, Jean-Marie Lehn (b 1939), a young Alsatian chemist who obtained his doctorate at the University of Strasbourg with Guy Ourisson (1926–2006) and then went to postdoc with R. B. Woodward (1917–1979) at Harvard, quickly rose to the firmament, at first with nmr studies of note.

His ambitions transcended nmr though. He aimed at becoming a chemist of the first rank. Which he did from combining the abstract and the concrete, categories of endeavor provided by his taste for epistemology and exploratory lab work, respectively. He read voraciously and in all directions the chemical literature. Given his outstanding talents, his was a meteoric rise.

In 1969, he published the very first papers in the chapter in his career that he would entitle supramolecular chemistry. They described synthesis of cryptands, molecules designed to encapsulate metallic ions (as cryptates). He presented this groundbreaking work at the Bürgenstock meeting in stereochemistry in 1970, May 3–9, in a plenary lecture I vividly recall.

Jumping ahead in years, Jean-Marie was awarded the Nobel Prize in chemistry in 1987, jointly with Donald Cram (1919–2001) and Charles J. Pedersen (1904–1989). The very first word in the abstract of his Nobel lecture, "supramolecular," deserves mention; if anything, because it conveys a scientifico-cultural atmosphere from the Sixties. Whereas the full phrase "supramolecular chemistry" is uniquely Lehn's, the adjective "supramolecular" was coined by a number of authors. Polymer chemists used it repeatedly in the late mid-to-1960s. In 1966, Albert L. Lehninger (1917–1986) wrote of the supramolecular organization of enzyme and membrane systems. George C. Oppenlander, an industrial chemist working in the R&D Center of the Hercules company in Wilmington, Delaware, published an article in *Science* (1966), asserting in its title "Supramolecular structure can be controlled, and influences mechanical properties." And Hans Selye (1907–1982)—we are indebted to him for the notion of stress—published in 1967 a book entitled *In Vivo: The Case for Supramolecular Biology*. These were, I submit, influential in Lehn's choise of supramolecular chemistry to denote the axis of his groundbreaking latter research during the Seventies.

An inspiration for Jean-Marie's work on cryptates came from biology, the transfer of ions across cell membranes, mediated by carrier molecules, such as ionophore antibiotics. I have already mentioned Lehninger—talked about at the time as a potential future Nobelist. Another scientist who advanced this field was Wilhelm Simon (1929–1992) at the Swiss Institute of Technology (ETH) in Zürich.

As R. B. Woodward (1917–1979) was about to retire from his Harvard professorship, the issue of his replacement came up. I recall vividly Ron Breslow's (1931–2017) joy and pride telling me—during a visit of mine at Columbia at the end of 1973—the solution devised by the committee he was chairing. Woodward would be succeeded by three young chemists in their early 30s, Kishi Yoshito (b 1937), Jeremy Knowles (1935–2008) and Jean-Marie Lehn (b 1937).

Jean-Marie indeed stayed awhile in Cambridge, Massachusetts during the spring semesters of 1972 and 1974. Ultimately though, after carefully weighing the pros and cons of either the US or France for his future career, he decided to stay put, in Strasbourg. In 1980, he was elected to a chair at Collège de France, in Paris, in replacement of Alain Horeau (1909–1992).

Discovery in 1954 by Watson and Crick of the DNA double helix influenced chemistry in depth: henceforth, many chemists saw the future of the discipline marked by biology-linked pursuits. A case from the Sixties was vitamin B_{12}. After Dorothy Crowfoot-Hodgkin (1910–1994) established its structure in 1955, R. B. Woodward (1917–1979) launched upon its synthesis, allying himself with Albert Eschenmoser (b 1925) at the ETH in Zürich for this monumental task. Others, such as Duilio Arigoni (1928–2020), also in the chemistry department at ETH, and Alan R. Battersby (1925–2018), who would become elected to a prestigious chair at the University of Cambridge, studied the biosynthesis of B_{12}. Indeed, biosynthetic studies were a sector bridging biology and chemistry in the Sixties. An off spring was a novel undertaking, viz. to model chemical reactions upon biological transformations. Ronald Breslow (1931–2017) at Columbia termed "biomimetic" this new feature of chemistry.

Why this adjective? Was it the relatively recent (1953) translation into English of the milestone book in literary criticism *Mimesis* by Erich Auerbach (1892–1957)? A much more likely source for the borrowing by chemists was the American inventor, engineer, and biophysicist Otto Herbert Schmitt (1913–1998): he lectured during the Sixties on biomimetic transforms, which his wife Esther, a biochemist, may have told Ron Breslow about. For instance, Schmitt urged study in 1963 of "biological phenomenology in the hope of gaining insight and inspiration for developing physical or composite bio-physical systems in the image of life" and used the term "biomimetic twice in the lecture."

Coworkers of Note

8

During my time at Princeton, I was close enough to graduate students to hear their typology of advisors, i.e., professors as leaders of research group. There were three categories, slave-drivers, operators and contractors. In addition, Paul Schleyer told me that, compared to the Princeton undergraduates, "graduate students (are) second-rate people." (authentic quote).

My intuition made me think and act very differently. Each person has a secret garden, the source of vast riches. To tap into it, one has to show utmost respect.

Such admittedly singular views are bolstered by the language. The etymology of the verb "to educate" shows its meaning to be "to bring up, to make grow." My belief is for a Ph.D. advisor's role to be analogous to a gardener's. It is akin to bringing an original body of work to fruition. In other words, those of Plato, it is the Socratic maieutics.

One has to remain patient. Some people need months, even years, to show their mettle. I recall a Princeton undergraduate in my group who spent all his time playing as a child in the hallway. He retained my respect. He was undergoing a spiritual crisis. He is now a leader in chemical science and a professor in a major university of the American Northeast.

I recall a graduate student, also in my group at Princeton, one of the smartest people I have ever come across, who spent the whole time reading novels. I recommended Alexandre Dumas to him; and he read the whole, huge opus. This extraordinary person is Sarangan Chari (b. 1946). He became a highly successful trader on Wall Street and still lives in Princeton. He works there in Michael Hecht's lab at the university: to quote from his self-description, "After many forays in industrial and academic labs, he wound up as an investment banker. Having seen the error of his ways, he is back in science trying his hand at Synthetic Biology. His current project involves deciphering the mechanism of auxotroph rescue by synthetic proteins. Outside of the lab, he is mostly concerned with ameliorating the effects of decades of sleep deprivation."

Other graduate students in my research group at Princeton included Arnold Speert (b. 1945), who became a college president; Arthur Greenberg (b. 1946) who not only

P. Laszlo, *A Life and Career in Chemistry*,
https://doi.org/10.1007/978-3-030-82393-1_8

became a professor of chemistry and a Dean at the University of New Hampshire, also an historian of chemistry and a bookwriter; William E. Frankle (1945?–1996), who became a pastor in New Paltz, in the Hudson River Valley; and Raymond Dratler, who after a post-doc in Jean-Marie Lehn's group in Strasbourg switched gears and became a dentist. I pride myself on such diversity of careers, being liberal rather than bossy may have had something to with it. All in all, I also take in that at least 20 of my former coworkers became university professors themselves.

Which brings up the topic of the shadow of the doctoral advisor. For the advisee, exposed to that influence not only daily but for several years, there are twin opposite temptations: submit to his profound influence, espouse his tastes and choices; or conversely find one's opposite path.

Looking back at the Ph.D. students who chose to work with me at Princeton, Arnie Speert's choice reminds me of the fleeting temptation I had to become a university president. Art Greenberg is a spiritual son—neither he nor I can deny it—for his historical interests, his collecting treasured books in the history of chemistry, his book-writing and, last but not least, the importance to him of his teaching, at which he is outstanding. Conversely, Bill Frankle and Raymond Dratler had to reject my influence and push away into their highly personal, even idiosyncratic careers. I am delighted that Raymond and I have remained very good friends.

I owed my call to the university of Liège, in French-speaking, Eastern Belgium, to a combination of friendship and luck. The good friend was Jacques Reisse (b 1936), professor of chemistry in Brussels. The circumstance was my having lectured there, at the end of May 1968 when the black flag of the anarchists still floated on the central building of the university.

Some years back, I read a piece on world tourism: the Belgian and the Swiss are the most adventurous, they go everywhere. I ascribe this spiritedness to the small size and pluricultural nature of their countries.

I was lucky to spend part of my career in Belgium, from 1970 until 1986, when I returned to France. We were several foreign academics teaching chemistry in Belgian universities during that period: Richard Martin (1914–1995), at the Free University in Brussels; Heinz Viehe (1929–2010) at the Catholic University in Leuven; Alain Krief (b 1942), at the University of Namur; and myself in Liège. Which testifies to both the commonsense and the hospitality of Belgians, who in addition entirely lack chauvinism.

Liège is in French-speaking Belgium or rather in what used to be a Walloon-speaking region—Walloon also being a Romance language. In 1970, when I arrived, only the elderly spoke Walloon. However, with such a recent and rather forced introduction of French, the population was still uneasy with their new language. This hybrid character extended to behavior, I found Walloons to be seemingly German-like in disciplined behavior, which overlaid a very French anarchistic streak!

I was able to build there a strong research group. The graduate students rewarded me with outstanding results. The first two, André Cornélis (b 1941–2021) and Jean Grandjean (b 1945), were transfers from the group of the retiring professor, Jean Baudrenghien (1897–1979), whom I replaced. They became loyal lieutenants and they helped me a great deal and over many years.

I shall mention only a few other of my coworkers, among the two dozens or so that I was lucky to supervise over the years: I taught in Liège from 1970 until 1986 as an *ordinarius* (full professor) when I answered the call from École polytechnique, in Palaiseau near Paris. I retained my lab in Liège though as an *extraordinarius* (visiting professor) until retirement in 1999. Two ladies greatly helped in serving as my secretaries, Nicole Dumont-Troisfontaines and Lucienne Souka, in succession.

Among my Liège coworkers, pride of place goes to Michel Jauquet, unfortunately killed in a car accident in 1973. The tragic death of this highly gifted young scientist put an end to a budding, although most promising career. He would certainly have become a professor himself. Our interaction was marked, both ways, by a combination of banter and respect. Michel, after he joined my group, espoused chemical physics. He had an eye for promising topics and areas, he was enthusiastic, he loved teaching: his was a grievous loss.

A worthy topic for historians of science is influence, from the field chosen for one's Ph.D. to scientific bandwagons and vogues—pathological episodes such as polywater and cold fusion included. Not only influence, the anxiety of influence as well, to refer to Harold Bloom's (1930–2019) remarkable study (1973) of writers at the time of Romanticism.

My autobiography confronts me with this paradox: as much as I strove to respect the autonomy of my coworkers, to eagerly expect their self-expression, yet as they became in turn research scientists on their own, with some the imprint left by our collaboration stamped them durably.

Is such sway only flattering to my ego? Did I fail in pushing them toward separate trajectories? Ought I to have been more forceful in guiding them away from my personal choices? Were these so luscious temptations as to have made them irresistible? How could I have helped these young adults to kill the father figure; should I have put more distance, even shown some nastiness?

These are my thoughts in retrospect and in surveying now their very real achievements once they were on their own. Before proceeding to mention by name some of the most memorable, an answer to the above paradox is needed: they worked very hard to deserve their Ph.D. degree, hence they were sorely tempted to maximize the rewards for their effort. Return on investment, seems to be the quick and dirty answer.

André Cornélis (b 1941–2021) after my formal retirement in 1999 was no longer active in scientific research. But he bore my imprint in teaching, involving himself in educational improvements at the University of Liège. They aimed at helping Belgian students make their first steps in chemistry.

Jean Grandjean (b 1945), who became an expert specialist in NMR methodologies, continued his momentum, using the 300 MHz spectrometer I had equiped our research group with. He was quite active during the first decade of this century, until his own retirement, pursuing topics we had earlier explored together, e.g. water at various clay interfaces; mostly in collaborations, with Polish scientists or with a polymer chemist, Professor Robert Jérôme (b 1942) at the University of Liège.

It is hard not to like Christian Detellier (b. 1949), he is one of the friendliest persons I have ever met. He was the passenger of the car in which Michel met his death in March 1973—this tragedy marked him but he did not let it ruin his life. Christian has a rebellious streak—a promising aspect for a scientist. Quick-tempered, he shows occasional flashes of anger or joy.

His postdoctoral appointment was with Henri Kagan (b 1930) at the university Paris XI-Orsay. Following which, after 20 months of civil service, he spent one year as replacement professor at the University of Ottawa, working in the lab of Howard Alper (b 1941). He emigrated then to Canada and spent his whole career at the University of Ottawa, a bilingual institution. There, because he is so obviously very genuine, he was elected chairman of the chemistry department. Following which, he was made into a Dean of the Sciences and a university academic vice-president and provost. Being an administrator, he had to deal with occasional hard times; such as a demagogue professor passing every student as a provocation to get rid of him and sue (he could get fired despite his tenure). As a Dean, he built a whole new complex for the biosciences and devised novel teaching programs, still active today.

Christian and his wife live in Quebec, at a relatively short drive from Ottawa in Ontario. They have two daughters, each with a fine education. The Detelliers's hobby is endearing, bird-watching—which takes them to remote parts of the planet, they have been ecologically-minded from birth, one construes.

While in my research group in Liège, Christian took an active part in the study of DNA quadruplexes with alkali metal ions. He informally founded a sub-research group with Freddy Delville—more on him below—and André Gerstmans. We owe them theoretical understanding of sodium-23 nmr, among other topics. While in Liège, Christian was infected with the urge to explore clay platelets and their properties. In Ottawa, his onerous administrative duties did not interfere with his continued research on clays such as kaolinite or halloysite. He has widely explored the potential of numerous kaolinite intercalates. Thus, I have several 'scientific grandsons,' one among the latest at the University of Yaoundé in Cameroon.

Alfred (Freddy) Delville had his education as a mathematician. He joined my research group in Liège when I realized our need for an applied mathematician to analyze some of our output data. He followed me to France when I was appointed a professor at *École polytechnique*, in Palaiseau. I obtained for him, at that time, a tenured position with CNRS. Then, he quit abruptly for the University of Orléans, lured there inter alia by a position in physics. But he remained faithful to the study of the physical chemistry and chemical physics of clay platelets suspended in a fluid: some of his publications of the 2010s deal with the identical topics and methods as those of 20 years before. Accordingly, I have also scientific grandsons, such as Patrice Porion, at the University of Orléans.

Am I more pleased or annoyed when former associates such as graduate students or postdocs follow in my footsteps? Maybe I should hold neither feeling. Most probably, this was the easiest course for them to follow, rather than them attempting to become active into another field.

Does that mean that by attaching their name to mine they obtained an informal license to practice science, provided that it be in the same field? This conjecture seems to touch on absurdity. However, we live in an Age of Specialization.

Has use of the Internet increased the perceived need to specialize? Perhaps, since we attach labels to ourselves when participating in a network, scientific ones included.

Has the financing of research with grants increased specialization? Most probably, since adjudication of research proposals is with panels made up of peers, within the same research area.

Is such continuity good or bad for science? The former: I vividly recall the envy of American colleagues, when confronted with the permanent positions offered within a system such as the French CNRS. In North America, they had to retrain anew successive graduate students and postdocs into experimental techniques and methodologies. The latter: it is obviously a recipe for sclerosis. Science thrives, for its advancement, on discontinuities, even more than on continuities.

Some brilliant people have a way of being that makes it seem easy: which "it?" Interacting with them. Their carrying about, whether in life or in professional life. And, above all, their achievements.

Balogh Maria (b 1949)—to name her in the Hungarian manner, same as the Japanese—is such a remarkable person. She came to my lab in Liège, in the early 1980s, probably upon Fischer Janos's, from Richter Gedeon, recommendation. Why? My conjecture is a strong desire for leading edge research, combined with a need for a little hard currency. The times were then very hard in Hungary.

There, she worked for 38 years (!) as a synthetic organic chemist in the laboratories of the Chinoin pharmaceutical company, which became part of the Sanofi group in 1993. Repetitive work, preparatory to patent applications.

She came to Liège rather regularly during the summers, often accompanied by her young son; once by her mother as well. She was an outstanding coworker, from her training, work habits and above all, personality.

Training: Maria received it at the Technical University in Budapest, the best institution in the country; Hungary has a great reputation for higher education. After a MS in Chemical Engineering in 1973, Maria received her Ph.D. in Theoretical Organic Chemistry in 1978. Experience: her years at Chinoin (1973–2011) had trained Maria to prepare a novel molecule on a weekly, if not daily basis.

Personality? Lovely. Calm, quiet and cheerful; Maria is a most charming person, very endearing and completely trustworthy. Her command of the English language is on a par with that of preparative chemistry. We published quite a few papers together. Also a book, *Organic Chemistry Using Clays* (1993), in the writing of which she proved equally gifted, in both the writing and a comprehensive mastery of the then available literature.

At this point, some of the readers may feel wary of reading about the uniformly stellar qualities of outstanding coworkers. It may be time to take a step aside, so to say. I suggest a reflective stance about the training of chemists, during the second half of the twentieth century to set these remarks in context.

In order to skip technical details entirely, I shall compare it with that of painters—during the Renaissance, say. In both cases, the would-be artist or scientist is apprenticed—read: prepares his/her career—to a master of the craft. In both cases, the first productions are unoriginal, they contribute to the master's output. When the apprentice in the workshop/laboratory has learned the craft and becomes a contributor to either the art or the science, there is an attempt at a personal style of some originality.

Obviously, numerous differences distinguish the twentieth century chemist from a Renaissance artist. The most obvious are in the transmission of knowledge and techniques: printed matter was the dominant medium in my time, from both reprints of scholarly articles and their photocopies—the Xerox photocopier was revolutionary. To a large extent, the reprint, rather than spoken instruction from one's supervisor was the main instructor in the craft.

Which reminds me of a joke cracked by Paul D. Bartlett (1907–1997) in a lecture, suggesting that *JACS* (*Journal of the American Chemical Society*), the then leading journal of the profession, ought to be printed on perforated pages, so that one could tear off and dump all but the experimental part of a paper.

Armel Stockis has remained a lifelong friend. But I am also hugely impressed by him as a scientist.

The work he did for his Ph. D. in Liège was a gem. It was inspired. It was brilliant. It is a joint contribution of which I remain intensely proud. That was in the early 1970s.

Following which, he spent a postdoctoral appointment with Roald Hoffmann at Cornell in 1977–78, performing quantum chemical calculations on metallacyclopentanes and other organometallic species. It was of top-notch quality too, Professor Hoffmann quoted Dr. Stockis by name in his Nobel lecture in Stockholm (1981).

During the 1980s, Dr. Stockis switched to an entirely different path, at first he directed the pharmacokinetics laboratory in a governmental entity in Belgium, with the mission to streamline applications by pharmaceutical companies for authorization to market new drugs. The period 1985–2000 saw him involved even more in clinical pharmacology. The next 17 years, the last in his career as a pharmacologist, he spent at UCB Pharma, the Belgian pharmaceutical company, overseeing development of new drugs, for treatment of epilepsy in particular.

Armel is a fascinating person. His main characteristic, I think, is his inwardness: he is a thoughtful, deeply reflective person, extremely smart, that does not interfere with the warmth of his personality, benevolent and caring. Had he chosen to do so, he would have made a superb university professor. His choice to stay in Belgium may have got in the way of a superlative academic career. I honor myself in having guided his first steps as a scientist.

Lionel Delaude, whose portrait will close this section, is now a professor of chemistry at the University of Liège. This scientist is very much a self-made man. I had little opportunity to influence his growth as a first-rate chemist, it happened during the years when my major involvement was at *École polytechnique* in

Palaiseau, France rather than in Liège, Belgium. His Ph. D. degree dates indeed from 1990.

Accordingly, Lionel basically self-developed at the University of Liège. His formative years coincided with the global rise of the Internet, Lionel is an expert navigator of the Web.

He is highly efficient, never submitted to me a report that was not a finished product, ready for publication. In personal interaction, he was a Buster Keaton-like man of few words who spoke up only to communicate, not just to make conversation.

Perhaps because I was such an absentee supervisor, after I had retired in 1999, Lionel closely associated himself with two Liège chemists involved with catalysis, Alfred (Freddy) Noels and Albert Demonceau. All three became professors at the University of Liège. One may regret that Lionel did not continue on his original trajectory—an emphasis on the original.

A word on citations: the masterpiece that was Stockis's doctorate upon publication in *JACS* in 1974 received 20 citations (ACS data), 24 (Google Scholar), i.e., a little above the standard for a *JACS* publication. The paper Lionel and I published in the *Journal of Organic Chemistry* in 1996 received 10 times as many: 232 (ACS), 395 (Google Scholar); brilliant work as well. The explanation for the difference is that the latter paper introduced a new tool, of widespread applicability. Also, in the meanwhile, from the 70s to the late 90s, there has been an influx of Second World scientists, from India, China, Maghreb and the Mideast, Gulf States.

Yes, I was very lucky to have attracted such a galaxy of brilliant coworkers who, as a rule, went on to build distinguished careers of their own. Did I give them as much as they gave me? This is not for me to say or even hint at!

We shall meet them again in the final chapter, to take stock of the enduring appeal of their work.

Remarkable Scientists

9

My appraisal may have value for its departure from conventional views. Some, István Hargittai (1941) for instance, are focused single-mindedly on Nobel prizewinners. Personally, I am not so keen on evaluations by fellow-chemists, they tend to be too narrow-minded and conventional.

What I appreciate predominantly in a member of my profession and makes me rank highly this person is the actual or perceived ability for ease in other fields as well. The example that comes to mind—a bit trite and boyish—is of the German chemist Hans Musso (1925–1988) confiding to me his past as a forger—never caught. He did it for the challenge: counterfeit postage stamps which he would then stick on envelopes mailed out together with genuine stamps.

Epistemological anarchism ("anything goes") is a trend in philosophy of science associated predominantly with Paul Feyerabend (1924–1994), who published a highly influential book in the mid-1970s. The above motto can be applied to a number of scientists who, during the second half of the last century, no longer felt constrained by a straitjacket of goals and methods in their chosen field and roamed freely.

I can think of quite a few in chemistry, some prestigious and others lesser known. Examples are Jean-François Biellmann (b. 1934), organic chemist turned also engraver, under the name Baltzen; Tjalling Charles Koopmans (1910–1985) who after a key contribution to quantum chemistry became a Nobel-prizewinning economist; Guy Ourisson (1926–2006), natural products chemist who went on to organic geochemistry or dermatochemistry, in addition to being the scientific father to Jean-Marie Lehn (b. 1939) and grandfather to Jean-Pierre Sauvage (b. 1944).

Such a libertarian spirit started flowering in the Sixties. It produced a Renaissance if not a Second Revolution in chemistry. Hence, it is not surprising if some of the most remarkable fellow scientists I came across, or was associated with, were such free spirits.

To me, admiration comes with much of its original meaning, a form of astonishment. Chapter 2 already communicated my admiration for some of the teachers I was so unbelievably lucky to have.

P. Laszlo, *A Life and Career in Chemistry*,
https://doi.org/10.1007/978-3-030-82393-1_9

Admiration calls also for emulation, the non-imitative wish to rise to the admired feat. The jaded phrase, role models, is called for here.

My first submission of a major stature physicist and chemist is of the Estonian Endel Lippmaa (1930–2015). I met him in Moscow, in 1979, we both intervened at a workshop on modern methodologies in nuclear magnetic resonance, organized by the Academy of Sciences of the USSR. He had to come down from St. Petersburg by the night train, with some of his coworkers, victims that they were of Russian ostracism: people from the Baltic Republics were then being submitted to what can only be called apartheid.

Upon meeting him, I was impressed by the brilliance and acuity of his mind. He struck me, in addition, as a highly cultured person. Fluent in English, he spoke of course Estonian and Russian. I surmise that he knew German as well. He had a truly Jeffersonian personality, a benefit to Estonia. But first, an historical reminder.

The Molotov–Ribbentrop pact of 23 August 1939 was ostensibly a non-aggression treaty. But secret protocols divided countries to be enslaved between Nazi Germany and Stalinist USSR. The Baltic states, Finland, Estonia, and Latvia would go to the Soviets. Poland would be split and shared. Lithuania would be annexed by Germany.

In the 1980s, Lippmaa went to work in German and American archives and uncovered copies of the secret protocols annotated in Stalin's hand. In June 1989 Lippmaa was among deputies of the three Baltic republics who challenged Mikhail Gorbachev (b 1931) at the Congress of People's Deputies in Moscow. In December 1989, an investigative commission resolved that the documents were genuine; that they conflicted with peace treaties concluded with the governments of the independent Baltic states in the 1920s and 1930s and that they ought to be voided.

As a consequence, in 1990–1991 Endel Lippmaa was instrumental in making his country an independent state again. He became Minister for the Eastern Affairs of the Republic of Estonia, in 1995–1996 Euro-Minister of Estonia.

Truly, a Jeffersonian personality. I owe him the resolve not to let attempted humiliations by others affect you.

I became acquainted with Max Tishler (1906–1989) towards the end of his most distinguished career, when he chaired the Department of Chemistry at Wesleyan University. I had an ongoing collaboration with Ted Weissberger (b. 1941) in that Department.

Max Tishler was a pharmacological chemist who led the Merck R&D from a small outfit to world class status. The list of the industrial productions to his credit is impressive: vitamin C, riboflavin, cortisone, miamin, pyridoxin, pantothenic acid, nicotinamide, methionine, threonine, and tryptophan. In just a few years, he increased the amount of available cortisone from milligrams to tons. Tishler also led a microbiological group that developed fermentation processes for actinomycin D, vitamin B_{12}, streptomycin, and penicillin.

Let me remind the readers of the origin of his calling. The influenza epidemic of 1918 found Max as a boy apprenticed to a pharmacist delivering drugs in his native Boston. The ill and dying were everywhere. Deeply touched by such sights, he opted for a career in health care.

When I knew him, in his 70s, Max was avuncular and gentlemanly, interested in learning about anything and clearly extremely smart. I vividly recall sitting next to him, together with the speakers at at least one of the Peter A. Leermakers symposia during the late 1970s held in Middletown, Connecticut, where Wesleyan is located; I had enough eminence in his eyes, which was flattering. Leermakers was a most promising faculty members at Wesleyan, who unfortunately died from a car accident in Mariposa, California in 1971.

To me, Max Tishler has remained an exemplary figure: an elderly scientist owes it to himself to remain active and productive.

Carl Djerassi (1923–2015) was a citizen of the world and a polymath. Born in Bulgaria, raised in Vienna, having then studied in the US prior to employment in Mexico City, he was fluent in a number of languages: this was the background for his striving for excellence in literature, in addition to being a luminary among chemists—renowned for the contraceptive pill but unsuccessful in his bid for a Nobel.

I have seldom encountered a more driven person. His sheer energy was phenomenal; which explains the name he made for himself in a number of endeavors, such as publishing poetry, novels and plays in addition to a most impressive output of scientific publications. I am in his debt for a week spent at the Swedish Polar Station, in Abisko, about 150 miles (200 km) north of the Arctic Circle (Fig. 9.1).

Also, I was fortunate to be invited several times at the Foundation, south of San Francisco, on a huge ranch he owned, established in memory of his daughter, an artist, who committed suicide (Figs. 9.2 and 9.3).

After he had sought me out, together we wrote a small play, entitled *NO*, meant for use in the classroom; it was published in 2003. His third wife, Diane Middlebrook (d. 2007), an outstanding scholar and literary critic, was absolutely charming. She was why he embraced literature with the ambition of excelling in it.

Carl was difficult to get along with, he tended to be, not only self-centered, but also pushy and argumentative.

Carl co-wrote a highly successful play, *Oxygen*, with Roald Hoffmann (b. 1937). It was staged at the turn of the century. Based on history, I believe its writing was influenced by the success of Tom Stoppard's (b 1937) plays such as *Travesties* (1974).

Roald is a close friend of mine. He received a Nobel prize in chemistry relatively early in his life (1981), which did not slow him down in his multi-pronged investigations of chemical species and behavior. Roald can also pride himself on his writings, poetry predominantly. A professor at Cornell, he had taken there classes by his colleague A. R. Ammons (1926–2001), an outstanding contemporary American poet.

Roald invited me to serve several visiting professorships at Cornell, in the late 1980s and early 2000s. We took advantage of these to write several joint essays, of which I remain proud. Roald is a most likable person, highly cultured, generous and hard-working in addition to being immensely talented.

Lest I give the impression of featuring only American chemists in this chapter, I'll turn now to French ones. I shall single out for praise, not only Guy Ourisson

Fig. 9.1   A house in Abisko, Lapland, Sweden (photograph by the author)

(1926–2006), to whom we owe directly the Nobel prizewinner, Jean-Marie Lehn (b. 1939) and indirectly another Nobel prizewinner, Jean-Pierre Sauvage (b. 1944), Lehn's coworker; also another Alsatian chemist, Charles Sadron (1902–1993), a hero in WWII who, a prisoner in the notorious Dora camp, sabotaged the V1/V2 rockets being built; and Jean Fréchet (b. 1944), with whom I coexisted at Cornell and whom I hold to be the foremost French chemist currently—he had the advantage of remaining uninfected by the French ills, having spent his whole career in Canada, the US and nowadays the Gulf States and China.

What I learned from Sadron was that biophysical chemistry would prevail in our future. And from Fréchet the importance of being personally well-organized.

French chemistry has indeed been plagued by liabilities, most obviously chauvinism, parochialism and monolingualism, CNRS with its tenured positions, the seeking of administrative power rather than quenching a thirst for knowledge. These criticisms come not only from insider knowledge, also on the advice given me by André Weil (1906–1998), when I was leaving Princeton for Liège: academic cooptation leads inexorably to mediocrity unless one chooses systematically colleagues who are better than oneself—which runs contrary to human nature.

Fig. 9.2 Djerassi Foundation, a view towards the ocean (photo by the author, January 2007)

Fig. 9.3 Djerassi Foundation, wildlife (photo by the author, January 2007)

Parochialism? In university appointments, all too often, instead of choosing the best person, a local candidate is appointed. In addition, top students go to *Grandes Écoles*, such as *École polytechnique* or *École normale supérieure*, rather than to universities.

Pursuit of power? French science suffers from *mandarinat*, i.e., the wielding of power by a few. Thus, CNRS for instance is structured into decision-making committees, each of which with its ruling chairperson, a man usually.

Insufficient funding is another source of weakness. I'll mention one area, the equipment of laboratories. All too often, it comes with the building of a new facility. Hence, renewal only after a longish period, 25 years or so. In addition, no monies are awarded for the all-too-necessary upkeep and upgrading.

To be at the leading edge of the quest for knowledge is an adventuresome and solitary position. France is a Mediterranean country with a closeknit family structure—think of Napoleon with his siblings—unfavorable to such a lonesome undertaking.

To mention a single factor among the weaknesses of French science, the worst perhaps is the absence of altruism—of providing for the common good instead of the selfish "every man for himself mentality." The above-mentioned Guy Ourisson was an exception. He had an unusual sense of duty and concerned himself repeatedly with the interests of the scientific community as a whole and accepted accordingly appointments to a number of leading positions. What I learned from Guy was that one's allegiance to the scientific community brings not only privileges, but duties as well.

Duilio Arigoni's (1928–2020) star was one of the brightest in the chemical firmament. A Swiss from Ticino, he had to prove himself at the Swiss Polytechnic School (ETH) in Zürich against prejudice, taunted as he was for just being a *macaroni*. He became a full professor there at age 39. A polyglot, he was fluent in half-a-dozen languages.

Conventional knowledge, during his entire career, marked him as a certain Nobel prizewinner. He shared in formulating the isoprene rule but not in the resulting Nobel prize. He continued though to work on the biosynthetic pathways of terpenes, discovering a new such sequence of metabolic reactions.

My recollections of him are predominantly of his brilliance. He allied a spectacularly quick and incisive intelligence with a vast culture, whether in fine arts, music, he held Mozart as his hero, literature, and history: a Renaissance man, the cliché is unavoidable.

As a small very mundane detail, I recall during some summer conferences we attended together and to which he was accompanied by his family, seeing his children wearing T-shirts in the colors of various American campuses. This was 10 or 15 years before fake outfits of that kind (predatory) began being sold and worn worldwide. I copied his habit and would bring back to Belgium college T-shirts for my daughters.

At the end of the 1960s, when I worked in the chemistry department at Princeton, we hosted splendid chemists as visiting speakers. Manfred Eigen (1927–2019),

Nobel prizewinner, made a huge impression on me, I have never met a better physical chemist, he was a master in that field.

Another visitor, who lavished on us a whole series of lectures, was Rowland Pettit (1927–1981). He was from the University of Texas, in Austin, where he had followed his mentor, Michael J. S. Dewar (1918–1997). He was originally from Australia, descending he told me, from French immigrants there (Roland Petit?). His work using metal carbonyls made such an impression on me that I became a follower. Two doctoral theses I supervised, those of Arnie Speert (1945) and Armel Stockis, were influenced by the enthusiasm that Pettit infected me with.

He was lost to science way too early. Like another great chemist whom I knew personally, Earl Muetterties (1927–1984), his smoking brought along the lung cancer that killed him.

What goes into the making of a successful scientist? The ingredients are numerous: a vivid curiosity; the ability to draw connections between seemingly unrelated items; an analytical mind; being good at numbers; observational skills; and so on and so forth. To these, I submit that being good at languages should be added. To be at least bilingual, if not a polyglot, is an asset, I submit. Why? Put in the most simplistic terms, because it enlarges the mind (Fig. 9.4).

Many of the Chinese are indeed bilingual, speaking one of the many regional languages and Mandarin, the lingua franca. This advantage is, I believe, one of the prime reasons for the fast scientific, technological and economic rise of their country in recent years. A trite comment is that Asians brought up on ideograms have an advantage in learning the kindred language, that of chemical formulas.

It is also an advantage to come from a small country in which being at least bi- if not trilingual is common. Switzerland is such a country and this chapter features a Swiss chemist, Duilio Arigoni. Another small country in which a significant portion of the population is bilingual is the Netherlands, where many speak both Dutch and English.

The Dutch colleague I shall feature here is Hepke (Heppie) Hogeveen (1935–2015). Born, if I remember correctly, to a Dutch father and a Polynesian mother, he obtained his Ph.D. at the University of Groningen, with Hans Wynberg as his supervisor. That alone was an omen for an unusual career, Wynberg (1922–2011) being such a flamboyant character and a daring chemist—echoing his bravery as an OSS agent parachuted into Austria during WWII.

I recall both Hogeveen and Wynberg visiting my office at Princeton, on separate occasions. Hogeveen to introduce himself and his then groundbreaking work; Wynberg, in particular, to try and recruit me as an author for his father-in-law's publishing house. Both were persons one does not easily forget. And both were doing outstanding work and publishing remarkable—and remarked—papers.

Heppie, prior to his appointment at the University of Groningen, worked at Shell Research in Amsterdam—then among the high-ranking industrial laboratories producing superb science, the Bell Telephone Laboratory, the DuPont Experimental Station in Wilmingon, Delaware or Varian Associates in Palo Alto, California being a few others. Heppie's work, on valence isomerization or on carbocations, was outstanding and turned him into a leader in those areas.

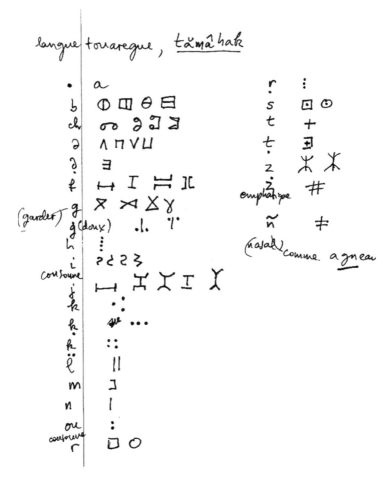

Fig. 9.4 A vivid curiosity? An example is my getting acquainted wih the Tamahaq alphabet in 1981 (a page from my notebook, photograph by the author)

My next and even more vivid recollections is from a Bürgenstock conference in stereochemistry, in May 1989, of Heppie's colleague at the University of Groningen, the American Richard M. Kellogg (b 1939) breaking to me the shattering news of Heppie's sudden disappearance. Dick Kellogg was devastated. Heppie had dropped totally out of sight. There was a report of him working in Piraeus as a stevedore.

Actually Heppie had suddenly and totally dropped out of both academic life and chemistry. He was no longer satisfied with that way of life. He needed a change. He sought another means of self-expression.

He found it. Or rather he found them. He underwent psychotherapy and obtained a degree in Gestalt therapy. In public he played the melodeon, "a type of button accordion on which the melody-side keyboard contains one or more rows of buttons, with each row producing the notes of a single diatonic scale."

But he was still dissatisfied with his life. Would he go back to chemistry? Not at all. With his wife, Ylil, he established himself in Lozère, a bucolic area in Central France. There, he became an artist, producing paintings and sculptures.

He was a gifted craftsman. As much as a chemist? I answer in the affirmative. His art is somewhat reminiscent of Paul Klee's (1879–1940), more accurately of Klee crossed with Vasarely (1906–1997). Also reminiscent of his past as a chemist. For instance, a sculpture shaped as a dodecahedron.

So what, a reader might shrug off this story. To jettison decades of labor and the huge talent demanded by a rise to the very top of academia commends respect; to work one's way to excellence in a radically different endeavor is also subject to esteem.

Remember: a majority of the scientists who ever lived were my contemporaries. Thus, I have encountered remarkable people who have honored our profession and excelled in it. However, the pace of change is such that their star shone only relatively briefly. History of chemistry seen from the inside may seem static to a participant-observer who has not moved along.

Instead, one ought to take at least occasionally another view. Historians are familiar with the Braudélian *longue durée*, the long haul in which centuries seem like mere years. I would liken the years of glorious success of an individual to the few instants in the limelight during which the rider in a rodeo hangs on for dear life before being thrown off the bucking wild horse.

Seminars and Conferences

10

I have probably presented at least 20 seminars yearly during my career. Hence, I won't be recalling each presentation precisely. But I can offer some generalizations.

The first is enthusiastic: the scientific community is a worldwide reality. You travel to the antipodes and meet colleagues there whom it is hard not to term brotherly. There is so much they quite often share: a taste for gourmet food and good wine, a love of classical music, familiarity with art and great literature, etc. In addition of course to related scientific pursuits (Fig. 10.1).

Whom should I cite here? The first who comes to mind is Alex Trifunac (b 1944). He and I met at a Gordon Conference—more on those later—in New Hampshire by a lake: during the conference, we went sailing together. I was already acquainted with his doctoral advisor, Gerhard L. Closs (1928–1992), a professor at the University of Chicago and a clever man responsible for milestones in physical organic chemistry; he and I had overlapped at the Institute of Molecular Science in Okazaki, Japan. Alex and his wife Marion, née Thurnauer (b 1945), who had also obtained her Ph.D. with Closs as her advisor, both worked at the Argonne National Laboratory, in the Chicago area. At the time I met them they had a passion for flying their small airplane. Alex took me for several rides in it. A vivid recollection is of my holding the controls for a few instants, and feeling in my hands the tremors of air motions. On such an excursion, Alex took me to the Taliesin compound built by Frank LLoyd Wright (1867–1959) in nearby Wisconsin. He did a low fly over and made a few circles so that I could feast my eye on that architectural gem.

I recall Allan L. Odell (1944-ca 1983) of the University of Auckland, in New Zealand. Our most friendly acquaintance began in 1979 and the around-the-world trip I took together with George C. and Linda Levy. The sheepskin coat I had purchased in New Zealand was stolen, together with all my belongings, at Gare du Nord in Paris as I was preparing to board a train to Liège on the very last leg of that trip. About half-a-year later, Allan came to visit us from New Zealand and stayed in our home in Plainevaux, near Liège: he carried with him from 20,000 km away, half the way around the globe, a replacement sheepskin coat. Can you imagine a more caring friendship?

© The Author(s), under exclusive license to Springer Nature Switzerland AG 2021
P. Laszlo, *A Life and Career in Chemistry*,
https://doi.org/10.1007/978-3-030-82393-1_10

Fig. 10.1 The antipodes: picture taken in Wellington, New Zealand, on April 23rd 2003. Left to right: Anne Pérez, PL, Valerie Laszlo his wife, Philippe Pérez, then scientific councillor at the French Embassy in New Zealand

Since I mentioned the University of Auckland, one of the successors of Allan in the chairmanship of the Department of Chemistry was Charmian J., O'Connor (b 1937). I remember a conversation with her, there in Auckland, during which she and I talked about ourselves in most mutually trusting manner. In recent years, she has worked relentlessly for the equality of male and female scientists. The Queen included her in her Honors List and Charmian is now a Dame.

Another friendship born from a seminar and treasured throughout most of my adult life is with Raquel Gonçalves (b 1948). When we first met she was a professor of chemistry at the University of Lisbon, and the editor of the *Bulletin of the Portuguese Chemical Society*. This was in the early Seventies by occasion of my having been invited to join a conference at the Gulbenkian Foundation. She and I shared a love of writing and of the history of chemistry. Henceforth, we remained in contact through correspondence and my occasional visits to Portugal for other seminars. One of them was at Guimarães, in the Minho Province. The invitation had come from Hernani Maia (b 1936) a peptide chemist who taught at the University of Minho. Raquel came to Guimarães to meet me. She and Hernani, who then knew one another only slightly, became better acquainted—and fell in love. She had never been married. His marriage was unhappy and his wife died not long after that first meeting. Raquel and Hernani married and she moved from Lisbon to Braga, where henceforth they would both teach and live.

I could add quite a few names to this roster of colleagues who became close friends, all over the world, and whom I became acquainted with from giving a seminar.

A characteristic of mine, when giving a lecture or presenting a seminar, is an aversion to blocking myself behind a lectern or another set spot, such as a desk behind which I am expected to sit. I try to keep moving, to walk up at least to the people seated in the front row, or even further—to appropriate bodily the entire room, in order to warrant continued attention from the listeners. Also, at the outset, I use this age-old trick of teachers all over, start with a low voice for a few instants and raise it only afterwards, for the audience to pay attention. Once it does so, I hide behind my presentation, doing my best to switch the attention of the listeners from myself to the data shown on the slides.

I have always done my best to avoid the behavior of a showman, intent on having the audience focus its attention on the speaker. The material I presented was all-important and ought to eclipse my personality.

Hence, I prepared it carefully. During my career, it had the form of the slide show. I have thus witnessed a number of technical changes. I have known large boards covered with chemical structures that were hung and shown in succession to accompany doctoral presentations—a leftover from the nineteenth century. I have prepared myself and used Polaroid slides; Kodak Instamatic slides that one fed into a circular tray—a so-called Carousel—for projection. The slides themselves were initially black and white, in the mid-70s white type on a blue background and then, especially with the advent of personal computers, full color at last. I have also suffered through the era of transparencies, shown with a so-called epidiascope, or retro-projector. Suffered? Yes, because the projection device made you motionless, chained to it.

What I have striven for, during all those (minor) technical changes, was to devise material that was informative, but also included a few light moments and wandered, at times, into the unexpected—to reward the patience of the audience. As for using a pointer, whether a real stick or a laser pen, this was not a dance I was ever a virtuoso of. And of course, my listeners had to contend with my French accent in the English language, a highly characteristic and even idiosyncratic feature of mine.

When based in Liège, I initiated a series of joint colloquia with three colleagues, organic chemists as well, from the University of Nancy in Eastern France. They were Bertrand Castro (1939–2014), Paul Caubère and Bernard Gross (1932–2014). The Liège-Nancy distance (less than 300 km) is small enough to make it wieldy using cars for the trips. Several such meetings occurred and they were highly enjoyable—either the group from Nancy coming to Liège or vice-versa.

They ended abruptly though—entirely by my fault. One time, my coworkers and I were in Nancy, and we made presentations of our ongoing work. As the last speaker, I gave a talk that I meant as a hoax. At the end, nearly everyone (me included) was smiling or laughing heartily. Bertrand Castro however swallowed it all. As soon as I had finished my outrageous spiel, he asked a question—a genuine scientific query.

When Bertrand realized that it was a hoax he was furious; and these Liège-Nancy exchanges came to an abrupt end. A pity! A year or two after that seminar in Nancy meant as a joke, I met a colleague from Nancy at a conference in Tokyo. In chatting with him, he told me that Bertrand was still furious at having been taken in. Not only these colloquia had come to an abrupt stop, so did any friendly interaction between Bertrand and I.

Fortunately, eventually his anger subsided. Quite a few years later, at the beginning of this century, we resumed our friendship. By then Bertrand and Brigitte his wife had moved to Montpellier. He had been hired as a director of research at the Sanofi pharmaceutical company. After retiring from that position, he pursued research at the University of Montpellier, in Odile Eisenstein's research group. About 2007, this brilliant outstanding scientist discovered he had a cancer of the blood—not a leukemia though. He battled it but finally died after seven years.

An episode in his life as a child impressed me considerably. His family was Jewish, they were hence in great danger during the German occupation of France in WWII. Bertrand's father, also a chemist, then directed a large factory in Chedde, in Passy at the entrance of the Chamonix Valley in the Haute-Savoie. One day, he led Bertrand and his brother—little boys both at the time—on mountain paths. And then directed his two sons to keep walking by themselves straight ahead: they thus reached the safety of the Swiss border, a few hundred meters away, where they were taken care of upon arrival.

But what are the benefits from presenting a seminar? In addition to the personal rewards, such as becoming acquainted with another scientific center and its leading officers, so to say, the local sights, etc., the scientific rewards accruing from the effort are well worth it. Also suggestions from colleagues about further work, both performing checks to the effect that the reported results are genuine and not an artefact; and new directions to be pursued henceforth.

But what about liabilities from presenting a seminar, are they only rosy aspects and no negatives? Of course, there are. The people attending to some extent come for the kill. A seminar is bullfight-like in this respect, the results of a poor presentation coupled with lousy results. Fortunately, I was spared such humiliation. It typically occurs during the discussion part, when colleagues challenge your interpretation. Discussion of results is a bulwark of science, shielding it from dogmatism, paternalistic authoritarianism and similar diseases affecting entrenched and sclerosed knowledge.

Often, during the discussion, one has to deal with a non-question coming from a heckler, someone who has come to present his—most of the time a man rather than a woman—views on some issue with little or no relationship to your presentation. He has come to use and abuse the pulpit provided by the discussion. Such intrusions can ruin a discussion.

Fortunately, Michel Serres (1930–2019) shared with me a trick to minimize the damage from such intrusions. You just loudly tell the intruder, "I could not have put it better than you just did. Next question please?" And you move on.

More frequent than the self-promoter, more annoying too, is the contrarian. He or she, in asking questions from the presenter, seeks to destroy your presentation. As a rule, this is an individual who has failed as a scientist for a variety of reasons: did not seize his chance in a timely manner; studied the wrong problems; or lacks the talent and imagination. The psychology is all too obvious, the individual's lack of success seeks a success for him or her. In tackling contrarians—I did not have to suffer one in each seminar, perhaps one in three—I always sought to maintain human respect for the person, to remain very patient, to answer every point and, most important, to keep my cool.

I should tell how Kurt Mislow—not at all a contrarian but a top scientist, a wise and clever mind, came to be such a master of the discussion, during the period following presentation of his results by a seminar presenter. Kurt would generally come up with an incisive question, often destructive, which no one else had thought of posing.

Often I sat next to him in Frick 309—the classroom in which organic chemistry seminars were held at Princeton. Thus I could observe his behavior. During the first five minutes or so of the presentation, Kurt gave it rapt attention. Following which he would stop from listening and close his eyes. During the rest of the presentation, Kurt would focus in his mind on the assumptions, both implicit and explicit, made by the presenter. Thus, he spent three-quarters of an hour having questioned these assumptions: no wonder his question or questions during the discussion period were so potentially devastating.

Some of my seminar lectures were at German universities, such as Berlin, Frankfurt, Freiburg, Göttingen, Hamburg, Heidelberg, Karlsruhe, Munich, ... A tradition there, prior to entering the lecture room packed with students, included the lecturer sharing tea and cookies with the local colleagues, for about half-an-hour. This whole group, or nearly so, then entered the lecture hall in abeyance of the academic quarter: the quarter-hour (15 minute) discrepancy between the defined start time for a lecture and the actual starting time.

I preferred small conferences much more than big ones. I'll deal first with the latter. These were for me, typically, the national meetings of the American Chemical Society. They were huge events, with many thousands of participants, held in large cities: New York, Boston, Chicago, Atlanta, San Francisco, Philadelphia, ... In such locations, big hotels and their huge ballrooms hosted the gatherings.

A vivid recollection stems from the ACS meeting in San Francisco in the spring of 1968 (March 31–April 5). This was the week of the assassination of Martin Luther King. Riots erupted in a number of American cities. This was also when (March 31) Lyndon B. Johnson (1908–1973) announced that he would not seek reelection as president of the US in the fall. An epochal week, hence of the meeting itself I hold few recollections.

Indeed, these ACS national meetings, at least to me were occasions for sightseeing in the cities where they were held. The number one recollection is of one such meeting, organized jointly with the Mexican Chemical Society, in Mexico City. This was in 1975. The meeting itself was deserted. By chance, I witnessed the presentation by H. C. Brown (1912–2004), the future Nobel laureate (1979). He was by himself with only the projectionist. If you wanted to meet a colleague or a friend, the

Fig. 10.2 On the way to an ACS meeting in San Francisco, Martine and I experienced a wilderness in Wyoming with a group from the Sierra Club. Postcard to my father (recto)

location had to be the pyramids at Teotihuacan, or the Archeology Museum in Mexico City—the most touristy spots.

When ACS national meetings were held in Boston or Chicago, I would spend much of my time in their outstanding local art museums. I had a fondness for Montreal, sometimes selected as a location for one of the North American meetings the ACS organized jointly with another national chemical society. Montreal is the largest French-speaking city in North America. Canada has always represented to me a nice blend of European and American values (Figs. 10.2 and 10.3).

I witnessed the inception of poster sessions in such conferences. The first I heard of them was from Ron Breslow during the 1973 GECO meeting in Locguénolé, Brittany. Probably from accompanying his wife Esther M. G. Breslow at a meeting of biochemists, he had seen his first posters. He waxed enthusiastically about them. Indeed, in those early times posters were still spontaneous undertakings. A young scientist, typically, would quickly scrawl some structures on a large piece of paper, typically with felt-tip markers of various colors, summarize in a few sentences a key result and perhaps complement a cartoon with a witty comment or quote. They were a joy to look at; they brought with them an element of childish play—a comment which I mean as a compliment rather than a denigration. However, within a few years, posters became regimented, stereotyped, dull and uninteresting: what a pity! They dried-up, all too rapidly, from inventiveness and creativity to productions from an implicit assembly line; a switch as unsurprising as it was, to my eyes and mind, unwelcome.

Fig. 10.3 On the way to an ACS meeting in San Francisco, Martine and I experienced a wilderness in Wyoming with a group from the Sierra Club. Postcard to my father (verso)

Not as frequently as with national meetings of the American Chemical Society, I would attend meetings of its British counterpart, the Royal Society of Chemistry. The contrasts were, unsurprisingly, a much smaller size and attendance; and a much reduced set of subdivisions, reflecting those within chemical science. Another contrast was the amenities, instead of the large American hotel rooms, there were tiny rooms in student housing. Also, these RSC meetings were held in delightful small cities, such as Exeter, Norwich, Swansea, York, … that held considerable charm.

When in the US, whether living there as during the second half of the Sixties, or having travelled for the primary purpose of testing the results of my research group, from Liège or Palaiseau in front of my American peers, I also attended from time to time a Gordon Conference.

Those were held in New England—in New Hampshire typically—in summertime and in preparatory schools then emptied of students. The formula—a little too formal, I think—was for the participants to consist of equivalent numbers of academic and industrial chemists. If it worked well in furthering mutual understanding, it worked poorly, since, as a rule, the industrial colleagues were not used to taking the floor, nor at being candid about their work—stifled as they were by an habit of secrecy The settings were, at least to my taste, truly delightful aspects of these Gordon Conferences: the hilly, meadowy and picturesque landscapes of rural New England. There were hikes to take on local trails in the afternoons following lunch, or boating on a lake; others among us would spend such leisure time in a gym, playing basketball or volleyball.

The atmosphere was often that of a return to college days. I thus recall a Gordon Conference, on the topic of Marine Natural Products—held not in New England, but in California. I was there as the lone (and lonely) exception. Nearly everyone else, whether from academia or industry, was a former member of a single research group. They relished their reunion and I remained an outcast among them.

And there were in my country, in France, the GECO summer meetings. The acronym stands for *Groupe d'Étude de Chimie Organique*. Guy Ourisson began them in 1959, on the model of the Gordon Conferences. I believe the very first one to have been held at Hotel de la Madeleine, in Sarlat, in the Dordogne area—Guy owned his childhood and summer home nearby, in Domme. My first experience of a GECO may also be from Sarlat—my memories of it are a little fuzzy. In any case, I attended at least half-a-dozen such meetings. When the organizers were academics, they were held in Spartan circumstances, the sessions in an elementary school, the meals and board in a modest vacation place—typically from the *Villages Vacances Familles* organization. When the organizers came from industry, the settings were a bit more lavish—typically a, hotel-restaurant with a reputation for gastronomy, graced also with a swimming pool. I found such material circumstances of little import, since France is graced with hundreds of different *terroirs*, each distinctive and to me endearing.

GECO meetings were international, to some extent, attended as a rule by a few Americans, Italians, Germans or British colleagues. With reference to truly international gatherings, I'll mention IUPAC (International Union of Pure and Applied Chemistry) conferences. I attended a few. The most memorable to me was held in Helsinki, in 1990.

It was my first trip to Finland. I was admirative of the splendid and well-kept Art Nouveau architecture of many buildings in Helsinki. Well-kept American cars from the Fifties were also to be seen cruising the streets. Open-air markets sold snap peas that the locals were fond of, to the extent that the area was littered with shells. Another recollection is of inhabitants bringing their rugs to the Baltic shore to clean them in seawater on wooden platforms specially built for that purpose.

Organic synthesis was the Helsinki conference theme. There was a roster of distinguished speakers, such as Ron Breslow (1931–2017), Clayton Heathcock (b 1936), Jean-Marie Lehn (b 1939), Steve Ley (b 1945), Martin Semmelhack, Gary Posner (1943–2018), Victor Snieckus (1937–2020), ... Many of these gave brilliant lectures that have left stellar recollections. Another recollection was biding my time. I was the very last speaker, the conference lasted for a week. Hence, I sweated out in unease through all the talks, worrying whether I would be up to the task (I did OK).

I recall an earlier IUPAC meeting, in Tokyo, in the mid-Eighties. My main recollection is of Ron Breslow theatrically holding a piece of paper with the most recent results from his lab at Columbia having been either telephoned or faxed to him. At that time, we all believed Ron to be an assured Nobel prizewinner. However, in subsequent months (1987), the news broke out that a graduate student in his research group had provided him with fraudulent results—which put paid to his hopes for THE prize—a shame since Ron was such an outstanding chemist.

Regarding international conferences, my favorites by far were the Bürgenstock conferences on stereochemistry. I attended a few. They were held at the Bürgenstock, a promontory overlooking the Lake of the Four Cantons in

Switzerland. It was a gorgeous location. It had three or four luxury hotels. Audrey Hepburn had her villa nearby. The idea—it made it possible for us scientists to enjoy high-class settings—was, at the reopening of this resort in May, to use us scientists to train the personnel starting or resuming work, prior to receiving wealthier guests. André Dreiding (1919–2013), a professor at the University of Zürich, organized the first conference in 1965. By the second conference, in 1966, his organic chemistry colleagues from the ETH (Swiss Polytechnic), also in Zürich, took over: Vlado Prelog (1906–1998), Albert Eschensmoser (b 1925), Jack Dunitz (b 1923) and Duilio Arigoni (1928–2020) who chaired this second edition.

I attended a number of Bürgenstock meetings: in 1966, where a few months before moving to Princeton I met Kurt Mislow (1923–2017) who then chaired the Chemistry Department and the Bürgenstock Declaration occurred; in 1970 when Bill Lipscomb (1919–2011), not yet a Nobel laureate, stole the show—whatever the topic, he knew it inside out; in 1973, when a frail, aged Robert Robinson (1986–1975) was the guest of honor; in 1978, when I presented a communication about the self-assembly of $5'$-GMP; the first week of May 1989, when I gave a plenary lecture on Organic Chemistry at Interfaces; in May 1991; there may have been yet other attendances of mine at the Bürgenstock.

I may have been a regular and I was not alone in this respect. Other faithful in attendance were from the Netherlands, Egbert Havinga (1909–1988); Johannes Dale from Norway; from Belgium Richard H. Martin (1914–1995) and Jacques Reisse (b 1936); from France, Jean Jacques (1917–2001), Alain Horeau (1909–1992), Henri Kagan (b 1930), Jean Mathieu (1917–2003) and Guy Ourisson (1926–2006); from the UK, Alan R. Battersby (1925–2018), W. D. Ollis (1924–1999), To attend these conferences regularly is readily explained, their intellectual eminence, the stimulation they imbued one with were matched only by the beauty of the site. Hence, the club-like character of these Bürgenstock meetings.

Recollections from these occasions? Reaching the railroad station in Lucerne, and then a short walk ahead getting on a ferry that toured the lake and would drop one below the Bürgenstock, one of the first stops, prior to riding an antique (1888) funicular railway up to the platform where the hotels stood; getting acquainted and befriending Roald Hoffmann (b 1937) upon his first attendance of the conference, introduced by our common friend Odile Eisenstein (b 1949); Robert B. Woodward's (1917–1979) 1966 lecture on the conservation of orbital symmetry; riding on a bus seated next to Martin Karplus (b 1930) and realizing his exceptional benevolence and kindness; chatting with Pierre Crabbé (1928–1987) during an afternoon cruise on the lake and realizing how insecure and pessimistic a person he was; Kurt Mislow, during an after-lecture discussion scolding Italian professor Fernando Montanari and starting his comments with "for your education, Professor Montanari, etc."; hiking in the afternoons in the Alpine meadows overlooking the hotels.

Personal Impact of Philosophy of Science 11

This chapter is in two parts. The first is a follow-up to Chap. 2: I mention in it philosophers who were my teachers and who kindled my interest in the subject. The second part returns to the topic of representation in chemistry, that Roald Hoffmann and I wrote about in an essay published some years back (1991) in *ACIE* (*Angewandte Chemistry International Edition in English*).

This is my first attempt at examining my link to philosophy, however strong. I shall recall initiation into the subject. Just as in my attraction to chemistry, I drew memorable mentors.

I will then proceed to philosophical issues met in my work as a chemist. And, in spite of reticence about being overtly critical, I won't resist a lament about the wasteland that philosophy of chemistry turned into in the last few decades. Thus a chapter starting by trumpeting hero-worship will end as a dirge.

First then, teachers. Let me preface this brief account with a reminder to readers: the French educational system includes a compulsory class of philosophy at the end of secondary school.

I took it in 1954 in Grenoble, at the Lycée Champollion and it was taught by Gaston Grua (1903–1955). Grua was an amazing person. He was a genuine philosopher in combining expertise in the history of philosophy—he published and edited a number of Leibniz manuscripts—, the teaching of philosophy and a life of modesty and even poverty; that of the common man. He was tolerant. He was benevolent. He was attentive to his students. It is worth repeating, he was an authentic philosopher, he was exemplary.

This was at Lycée Champollion in Grenoble, his penultimate posting prior to appointment at the University of Rennes to succeed Jean Anglès d'Auriac (1902–1954).

Grua graduated from *École normale supérieure* in 1928, with the *agrégation* licensing competitive examination. A classmate of his who remained a lifelong friend was the better known and longer lived Vladimir Jankélévitch (1903–1985). When I knew him in Grenoble, Grua led a Gandhi-like life, in a modest dwelling and

less than flashy clothes. He was a pious Catholic, a very gentle soul and the epitome
of philosophy.

Having already read on my own Kant's *Critiques* and Spinoza's *Ethics*, I was
very much interested in philosophy. Grua, as a teacher, showed the few of us who
were interested—half-a-dozen at most—how philosophy can discuss any issue and
strip it bare of all but the essentials. He never once, unfortunately, referred to his idol,
Leibniz: for his doctoral dissertation, a few years earlier, Grua had brought out a
critical edition in French of the *Theodicy* and another of so far unpublished
manuscripts he had studied and copied in Hannover, prior to the onset of World
War II.

My next encounter with philosophy was during my years (1955–58) of the NSE
classe préparatoire at the Lycée Saint-Louis in Paris. The teacher was another
graduate from *École normale supérieure*, Jean-Toussaint Desanti (1914–2002). He
was kind and gentle, as well as very smart.

At the time, we knew him to be infamous, as a leading intellectual who after
joining the French Communist Party, became in public and print a leading Stalinist
apologist—to the extent of applauding Stalin as a genius thinker and defending the
scandalous Trofim D. Lyssenko (1898–1976) self-appointed geneticist. However,
Desanti saved his soul by resigning his membership in 1956, after the Budapest
uprising; and turning back to epistemological pursuits into the foundations of
mathematics.

What did I get from Desanti? Not very much, to be honest. However, he had us
read with him La Mettrie's *L'homme-machine* (1747), a highly interesting
eighteenth-century pamphlet, which I would not have otherwise encountered. Grua
and Desanti both embodied to me the philosopher as someone who excels at making
students express and discuss their opinions, while respecting them even if far-fetched
and outrageous, thus the spirit of tolerance.

At the suggestion of two of my literary friends at École normale supérieure, Pierre
Macherey (b. 1938) and Jean-Paul Pittion (b. 1937), I attended the entire course
Deleuze taught at the Sorbonne in 1959–60. The subject was "*L'état de nature chez
Rousseau.*" It was attended by a small crowd led by the *normaliens*.

What gave Gilles Deleuze (1925–1995) such a commanding presence? Neither
tall nor bulky, he was of medium height. Nevertheless, he was charismatic. It was the
combination of his being handsome, well-dressed, endowed with piercing eyes and a
highly articulate voice. He was a consummate public speaker. Plus, a very gifted
teacher in spite of having ranked only eighth at the agrégation competitive examination in 1947.

This was philosophy that was both original and brand-new. It both resurrected
and transmitted thoughts from a few centuries ago, yes, but still very much alive and
relevant—to use the jaded phrase. With Deleuze the quest for knowledge became an
adventure. It was fascinating to listen to him present it. He was a spellbinding
lecturer, truly superb. He knew his chosen topic inside-out. He combined erudition
in the history of philosophy with love of literature and familiarity with literary
criticism. His admiration of Bergson (1859–1941) to me was an extra plus.

It was fascinating, listening to Deleuze, to find oneself in the company of this trio of great public intellectuals in the eighteenth century: not only Rousseau (1712–1778), a somewhat disreputable ruffian; but also Voltaire (1694–1778), whom I always found too tart for my taste; and Diderot (1713–1784), whom I totally admired.

Allow me a small digression here. In the summer of 1958, I had an experience duplicating one of Rousseau's. On October 24 1776, while in Ménilmontant, he was knocked over by a large dog. Rousseau lost consciousness. When he awoke, he became entranced by the beauty of the world and the sheer joy of being alive. In July 1958, I fell about 20 m (70 ft) in the Sept Laux range of the Prealps near my home in Grenoble. I did not lose consciousness but afterwards this near-death experience was also one of complete ecstasy.

But back to Deleuze and philosophy. In the Sixties, he together with Michel Serres (1930–2019) and Michel Foucault (1926–1984) showed that philosophy could address any topic whatsoever. This was a breath of fresh air after the total dominance of French intellectual life by Jean-Paul Sartre (1905–1980) and his existentialism. Deleuze could take-up Sacher Masoch, Serres could do likewise with the Jules Verne novels or the Tintin comics and Foucault the paintings of Velazquez, in addition to the history of science for these last two, both former students of Georges Canguilhem (1904–1995).

To me, becoming acquainted with Deleuze through his brilliant lectures brought home the cleansing notion of philosophy as new and exciting.

But in what way did I make it relevant to my practice as a scientist? About a year after auditing Deleuze's classes, heeding my call as a chemist I entered Edgar Lederer's (1908–1988) laboratory. This was as he transferred his operation from the *Institut de biologie-physicochimique* on the Montagne Ste. Geneviève, in Paris, to a brand-new CNRS building in Gif-sur-Yvette, in distant southern suburbia. Lederer studied plant natural products in the terpenes family.

My very first assignment was to apply nuclear magnetic resonance (nmr) spectroscopy to the structure of the rimuene diterpene. Rimuene is an extract from the tall New Zealand *rimu* trees that provided masts for the giant clipper sailboats, that roamed oceans in the nineteenth century.

My nmr spectrum unambiguously showed that the structure announced, just a few weeks before, by Ernest Wenkert (1925–2014) in the *Journal of the American Chemical Society* (*JACS*) was wrong. This finding, unwelcome to Lederer I note in passing—introduced me to the critical dimension of science, its life and blood.

Science thrives on even polemical discussions, on the steady attempt at refuting anything, whether an assertion, a conjecture or, as with the rimuene structure, a construction based on hard facts. This brush with the fundamental negative aspect of the quest for knowledge did not come to me as a surprise, nourished that I had been on Bachelard (1884–1962), on his *Philosophie du non* book (1940) especially.

In the aftermath and during my entire career, I knew in my very bones that the quest for knowledge thrives from challenging any statement or report. This was my experience from both seminars and publishing in peer-reviewed journals. Skepticism vivifies science—it keeps it alive—in like manner as strong acid dissolves the wax and makes the picture reveal itself in an etching.

Given my fascination with philosophy, why did I become a chemist rather than a philosopher? Answers to this question obviously belong in this memoir.

There was my interest—an understatement—in material changes. Among such transformations that transfixed me as a boy and a teenager, those of clay took first place. The plastic material could be scooped out from a clay pit. One would then mold it and, after drying in the sun, it would turn into durable little artefacts or sculptures.

My internship with Jeanne and Norbert Pierlot, the professional potters, taught me use of the potter's wheel to turn unformed clay into a pot, typically symmetrical about an axis. This was a first-hand experience into the plasticity of a material. For me, it reached back to the prehistory of mankind, when this craft was born, in conjunction with quite a few others through which hominids started imposing their mastery upon nature.

From the pit to the wheel, and from the wheel to the kiln: this second step is determining. Firing the pot turns it from a fragile construct into a most durable object, often a utensil. Archeologists have unearthed ceramic pieces dating back millennia; for instance, Jomon pottery from Japan dates back-16,000 years. A small figurine of a woman is the earliest known object made of fired clay, from almost 30,000 years ago.

What happens within the kiln during the long hours of the exposure of the dried pot to high temperature, above 1200 C for the stoneware is rather mysterious Norbert Pierlot kept telling us. He thus ignited a desire on my part for elucidation.

Another source for my interest in material changes brought about by heating—as good a definition of chemistry as others—was cooking. From very early on, I was fascinated by this art, of turning meats, fruit and vegetables into delicious foods, in particular the evolving of mouth-watering scents. Olfaction is amazing, humans can discriminate more than 1 trillion different smells.

Pottery is a craft that can be turned into an art. Likewise with cooking, chefs prepare gastronomical dishes. I was also impressed, as already mentioned, by the art of chemograms: a few drops of different chemicals on a piece of filter paper—same as used in the coffee machines present in many households—can produce a superb set of colors.

I was also attracted, during my teen years, by the art of photography, i.e., by chemical processes in the evolvement of art pieces. Need I say more? My becoming a chemist had an inevitability, brought about by its distant origins as a craft, which could be turned into an art.

In addition, my intuition was reminding me, as a young adult, *il n'y a pas de sot métier* (no stupid job exists). Hence, I did not care if some people undervalued chemistry compared to physics; and if biology appeared to present a more glorious future: I would become a chemist.

To return to philosophy as a temptation of mine, I partly fulfilled it, by teaching for several years at the University of Liège an introductory course for the students enrolled in the Faculty of Sciences.

I turn now to use of models for analyzing data. This aspect of the scientific activity motivates philosophers of science, rightly so. Not being scientists, they view it from the outside. Thus, they tend to make comparisons, likening modelling to make-believe play by children or to literary fiction-writing. Both are rather misleading.

To us scientists, a model is more like a tool in a toolbox. There are all kinds of models, from kits of molecular models to be handled and assembled into representations of molecules to sets of mathematical expressions. But all models have predicates, they all rely on assumptions. Assumptions, not true-or-false logical propositions, as some philosophers construe, mistakenly I think.

The most useful way to consider models, I submit, is from the etymology of the word. "Model" and "mold" are cognates. To model some data is to put it into a mold—and then look at it anew, molded. The model/mold doublet is not unique to English, it occurs in many other languages, for instance *modèle/moule* in French. Both derive from the Latin *modus*, a term used initially in measuring the area of a field. That the modeling activity is related to a measuring one makes sense, science demands both qualitative and quantitative thinking. A word derived from *modus* is *modulus*, i.e., a small measure: both "model" and "mold" derive from *modulus*.

A model is thus a kind of measuring device, a contraption applied to a phenomenon under study. It is a projection a scientist applies or attempts to apply—to the real world.

An important distinction, of a sociological nature, is between ready-made and self-designed models. The former are found in textbooks and in the literature. Most often, they result from a theoretical treatment of some generality. Some chemical examples from my practice were the transition state, acidity functions and the Woodward-Hoffmann rules for electrocyclic reactions.

The above distinction between ready-made and self-designed models was impressed upon me by an episode during the Sixties. We studied in my lab how the camphor molecule interacts in the liquid state with solvent molecules having benzene rings. A colleague studied the identical phenomenon. He and the graduate student he supervised chose as a model one-to-one complex formation between camphor A and the B solvent molecules, creating a long-lived AB entity. Thus, the (ready-made) model they were using prejudged the very nature of the interaction.

We made no such assumption in trying to account for the data we had assembled. What we found was that, instead of a long-lived AB complex formation, we dealt with a sequence of fleeting AB collisions, each lasting less than the time required for the camphor solute to turn around its inertial axes by a minute angle. We had to construct a model of our own design rather than resorting to an existing model from the literature.

To put it briefly, the colleague made an incorrect picture of the phenomenon by constraining it on the Procrustean bed of an existing model. Thus, I was not surprised when the same person was swept by the polywater wave and started studying this entity—an artefact, a kind of a silica gel rather than an anomalous form of water, as now everyone knows. There might be a related danger in sociological (external) accounts of the scientific activity, as compared to intellectual (internal) ones.

The take-home lesson is for a scientist to keep working at a distance from the crowd!

My career saw the birth and fast growth of computational chemistry. A burst! Indeed, I can recall vividly my using, as a student, archaic mechanical calculators. The present day started for me with purchase in the Seventies of one of the first commercial pocket calculators, the programmable Hewlett-Packard HP-25. On lecturing trips, I would use it to analyze the data provided by my coworkers, in anticipation of writing-up our results for publication. In doing so, I'd resort to conventional models, or I would assemble my own models.

The end of the century brought the big change. It is still too early to tell if it is a progress or a regression. That graduate students could draw henceforth on devices such as Monte Carlo-type simulations by using powerful numbers-churners, artificial intelligence as well, has changed the rules of the game. Chemistry from cottage industry becomes a factory production: is this for the better or for the worse? My instinct tells me the latter, more and more chemists turn themselves into efficient accountants.

I owe my education in theoretical chemistry to attendance of several summer schools on the subject: that led by Charles Coulson (1910–1974) in Oxford during the summer of 1961; during the following summers, those devised by Raymond Daudel (1920–2006) and Bernard Pullman (1919–1996), held in Menton. These summer schools featured talks by leaders in the discipline. There was a visible tension between proponents of extensive calculation and advocates of semi-empirical methodologies. I was intuitively favorable to the latter, as advocated for instance by John A. Pople (1925–2004) or Michael J. S. Dewar (1918–1997).

My take-home lesson, my taste as well, were and remain to privilege a qualitative understanding of one's data. My love for teaching furthered that imperative. Is it now old-fashioned into extinction? That would be deplorable. I feel welded to the phrase "to gain an understanding."

To understand a phenomenon, if one thinks of it, is a feat, an amazing and paradoxical achievement: the mind has conquered and mastered some bits and pieces from the real, material world. Alternatively, does one delude oneself about gaining *any* understanding? Does the world out there remain so much opaque to our manipulations and proddings that we are better off treating it as a black box, an input-output transformer, only ascertaining through numerical simulations that we can figure out its output with some accuracy?

A vivid recollection of bridging the microscopic and the everyday macroscopic dates to the mid-1960s. I was using an nmr spectrometer to monitor internal motions in molecules, known as Vorländer adducts: these consist in two rotors attached separately to a single carbon atom.

As I changed the temperature in the test tube, cooling it with an acetone-dry ice bath or heating it with hot air, their motions would slow down or accelerate. I was able to measure, when slow the numbers of rotations per minute, when fast these numbers per second. I was transfixed with the experience of monitoring such changes at close range: I had the illusion of being nested on top of the molecule (s), watching the wings flip over more or less rapidly. It was eerie. Definitely, a case of naive realism.

An analogy would compare these molecules and a windmill. Analogies populate the scientific mind. I can offer another case: in the early 1960s, I was investigating the effect of ring size on the pairwise interaction of atoms. The more you distanced from one another the atoms in a pair, the weaker their coupling became, as the ring they were part became increasingly strained.

Thinking about this phenomenon brought to my mind, in the most natural manner, having watched as a youngster archery tournaments in England. The molecular ring strain reminded me of the bending of a bow to tense it, prior to releasing an arrow flying toward uts target.

An analogy of this sort is to be both welcomed and resisted. It is beneficial, in that it connects the two worlds, that of events and activities—such as sports—in the everyday; and that of atomic motions in molecules, at the microscopic or nanoscopic scale. Such analogies were behind the conception and development of molecular machines by my friends Fraser Stoddart (b 1942) and Jean-Pierre Sauvage (b 1944), which shows their pragmatic utility.

Yet, in another sense, such microscopic-macroscopic analogies should be put aside for the simple reason that atoms and molecules also show behaviors foreign to the common experience: all kinds of quantum effects, jumps, exchanges, tunneling and the like.

The mature scientist is trained into a compromise between both these viewpoints. The former is useful as a teaching ploy in a seminar or a classroom. The latter is the—surely puritanical—attitude of respect for the strangeness of the microscopic world.

Which brings up the meaning of understanding: it is to refrain from intuitive descriptions, however tempting. Surely, they help the discovery process. But one has to ignore them and resort instead afterwards to a theoretical language. Despite its opacity to the common man, it provides the only adequate description of the microscopic reality.

Microscopic-macroscopic analogies, between molecules and real life—in its usual meaning—have other merits than the didactic. To return to the example of the effect of ring size on the distance of a pair of atoms and to take up again the analogy from archery, just as releasing the bend of the bow to send an arrow on its way, release from ring strain can propel many a chemical transformation.

While attempting to fertilize my scientific work with insights from philosophy, I continued to enjoy delightful times from listening to philosophers. There was, in the late 60s, a lecture series at Princeton presented by Gilbert Ryle (1900–1976). It was extremely enjoyable; educational as well, from his well-argued criticisms of Descartes.

Gregory Vlastos (1907–1991) was then a member of the philosophy department at Princeton. I did not meet him or hear him lecture at the time. Only at the end of the century was I privileged to do so: I sat in on a class at Cornell, to which he had moved and where I was a visiting professor.

Fig. 11.1 The naive urge to term new or novel one's ideas ought to be pitted against the permanence in one's environment. A fruit stall in Naples (May 2007, photograph by the author)

It was wonderful, it was endearing, it was charming! Vlastos was talking about Socrates, whose views he was utterly familiar with as if he had been an old buddy. The intimate familiarity was impressive. It was comparable—and this is the adequate manner to conclude this chapter—to my own intimacy with Vorländer adducts (Fig. 11.1).

Joy and Pride of Writing

12

Publication: the publish or perish alternative; Bruno Latour's (b. 1947) insight, a lab's main output is papers in scholarly journals (Fig. 12.1).

I was captivated by publication, from writing and submitting articles, refereeing manuscripts by other chemists, being an editor in several journals, of collective monographs as well, in addition to authoring a number of learned books and popularizations.

An orphan at age 7 I acquired and my widowed father encouraged a yearning to emulate my lost mother. She was exceptionally well-read and wrote poetry herself. From an early age I had resolved to become a writer. Having become a scientist, I just welded these two components—mutually supportive rather than antagonistic, as I found out.

During a career nearly 40 years-long, I witnessed changes in the process. Like most historical changes they amount to a zero-sum game, some desirable, others simply bad.

To start with the positive, the infamous Robert Maxwell (1923–1991) founded and headed Pergamon Press until his demise. He sped up chemical publications with his initiative of camera-ready copy. Hence publication in *Tetrahedron Letters*, a weekly journal *he* started in 1959, trailed results by only a few weeks. *Tetrahedron Letters* published organic chemistry, in the Sixties a vibrant sub-discipline of chemistry. It featured natural products, the cognate areas of medicinal chemistry and pharmacology, budding organometallics, mechanisms of organic reactions, etc.

During the Golden Sixties, book publishing saw the meteoric rise of the New York-based W. A. Benjamin imprint. Richard Feynman (1918–1988) owed this science publisher his near-instant star status. Many other physical scientists, including an impressive proportion of future Nobel prizewinners were also on the Benjamin list. Bill Benjamin took full advantage of the Paperback Revolution and rise of research universities. His was a very American story, in both daring and moral values. His main, radical insight was to connect scientific advances in the making with freshmen college classes. Such use of a book rekindled the habit of the faithful,

© The Author(s), under exclusive license to Springer Nature Switzerland AG 2021
P. Laszlo, *A Life and Career in Chemistry*,
https://doi.org/10.1007/978-3-030-82393-1_12

Fig. 12.1 Front cover of
Citrus, University of Chicago
Press, 2007

whether Jewish or Protestant, to anchor their belief directly in Bible reading. With science as the new religion: straight from the lab in like manner as direct from God.

Other ingredients of the Benjamin resounding success was was that he was the nephew of the McGraw-Hill CEO Curtis G. Benjamin (1901–1983); offering his authors a retreat at Martha's Vineyard to coax them into completion; and a board of directors of very influential top American scientists, such as Konrad Bloch (1912–2000) and Donald F. Hornig (1920–2013). Last but not least, having as an advisor John D. Roberts (1918–2016), a Caltech professor of chemistry with a gift for the written word and an amazing intuition for eminence in science at the budding stage. Instead of an uninteresting committee-chosen list of authors, near-exclusive reliance by Bill Benjamin on Jack Roberts's advice produced a scintillating roster of authors. During my Princeton years at the end of the Sixties, I used a couple of W.A. Benjamin books in classes I taught: Ron Breslow's (1931–2017) on *Organic Reaction Mechanisms* and my senior colleagues Kurt Mislow's (1923–2017) *Introduction to Stereochemistry*.

I taught at Princeton a graduate course on determination of molecular structure using physical methods. Peter J. Stang (b. 1941), then a post-doc in Paul Schleyer's (1930–2014) group, came one day to me suggesting I write a book based on it. My response was to draft him as a co-author. We entitled it *Organic Spectroscopy* and Harper and Row, another publisher in New York, brought it out in 1971, in a collection supervised by Stuart A. Rice (b. 1932). It was my very first book; quite a few others would follow.

However, Harper and Row at that point discontinued their scientific publishing. *Organic Spectroscopy* received little promotion if any. In spite of the lack of a commercial success, this book met with esteem on the part of a number of colleagues. And Peter Stang became a lifelong friend.

Upon my return to Europe, at the beginning of 1970 to take-up a professorship at the University of Liège, in French-speaking Belgium, I had already a scientific publisher in Paris. Named Hermann, it belonged to Pierre Berès (1913–2008), a flamboyant figure.

Berès, illegitimate son of a Russian prince, was raised in Paris. He attended the prestigious Lycée Louis-le-Grand for literary study. As a teenager, he was already dealing in rare books. Aged 18, he opened a bookshop on Avenue de Friedland. The German writer Ernst Jünger (1895–1998) visited regularly during the war years. Pierre Berès, who compiled and published catalogs of great bibliographic value, had other interests too. He was an art collector of note, with a keen eye to abstract expressionists such as Olivier Debré (1920–1999) and Pierre Soulages (b 1919).

In 1956, he bought the Hermann publishing house. He turned it into the Parisian counterpart of W. A. Benjamin, in New York. In particular, he pursued publication (until 1980) of the revolutionary and extremely influential *Éléments de mathématiques* by Nicolas Bourbaki—the pseudonym for a group of mathematicians intent upon formalization.

A few years later, Berès and I became acquainted. The *Times Literary Supplement (TLS)* had published a scathing article on "French Clarity" and I wrote a letter to the editor (issue of May 25 1962) to protest this denigration. The *TLS* published it. As I was preparing to leave France to teach at Princeton, he sought me out.

We became friends and he hired me as a consultant to advise about scientific manuscripts mostly. Michel Foucault (1926–1984) was another of his friends/ consultants. Foucault started the "*Savoirs*" collection at Hermann (having met him at the Berèses I attended his inaugural lecture at Collège de France).

My acquaintance and collaboration with Pierre Berès lasted until the 1980s and early 1990s. The first of my books that Hermann brought out was my translation of *Organic Spectroscopy* that came out in 1972. Followed a series of four textbooks, *Leçons de chimie*—thus entitled, not a little arrogantly on my part, under the influence of Richard P. Feynman's (1918–1988) *Lectures on Physics* (1963)—they were published from 1974 until 1984.

I was then a faculty member at the University of Liège and I ran a research group there. Belgian colleagues were quick to note my proclivities in publishing. They entrusted me with the editorship of a scholarly journal, *Bulletin des sociétés chimiques belges*. This was my first experience at being an editor, it lasted about 5 years during the 1970s.

To benefit from Belgian hospitality was special—and precious. There were, at that time, three or four of us foreigners as professors of chemistry in Belgian universities: Richard Martin (1914–1995), a Swiss, in Brussels; Heinz Viehe (1929–2010), a German, in Louvain-la-Neuve; and Alain Krief (b. 1942), another Frenchman, in Namur. An instance of the generosity of our hosts was that—at a time of great tension between Dutch- and French-speakers—they allowed me to use French in all professional meetings that I attended.

Back to my editing the *Bulletin*. It was straightforward. Chemistry in Belgium was at a more-than-decent level. It compared favorably with France, while not up to the lofty levels of Switzerland. To find referees for the papers submitted I was not shy and I drew upon the worldwide chemical community. I recall submission of a paper on electronegativity, which I sent to Linus Pauling (1901–1994) for review. It was negative. The author responded by questioning Pauling's competence!

My job entailed—I found out the hard way—setting many submissions into decent English. A yet more delicate aspect was to encourage Flemish colleagues to publish their work: at that time the quality of their research was quickly catching up with that by French speakers. A nice recollection is of a visit in Leuven to the shop printing the *Bulletin* where I was presented with a superb coffee-table book, *Notre Brueghel*, from 1969.

Meanwhile in France, Lionel Salem (b. 1937) began *Nouveau Journal de Chimie* in 1977. Lionel, like myself a graduate of the NSE *classe préparatoire* at the Parisian Lycée Saint-Louis, was another good friend. I volunteered to write a column on scientific history. I entitled it "*Petite chronique archéologique*," in obvious homage to Foucault's *Archéologie du savoir* book. A dozen such texts appeared during the 1980–1987 period.

A few years later, Olivier Postel-Vinay (b. 1948), a former foreign correspondent at *Le Monde*, became in 1995 editor of *La Recherche*, a monthly magazine of science popularization. He appointed me as a columnist on chemistry. I contributed more than 30 such entries at the end of the century, under the general title, "*La nature et l'artifice*" (the natural and the artificial). I loved that experience, especially meeting with fellow-columnists such as the neuropsychiatrist Boris Cyrulnik (b. 1937), the demographer Hervé Le Bras (b. 1943), the molecular biologist Antoine Danchin (b. 1944), the physicist Etienne Klein (b. 1958), ...

All such pieces were in a short format, a favorite of mine. As a young adult, my readings of recent and contemporary French literature opted for brief texts: the *Propos* by the philosopher Alain (1868–1951), prose poems by Francis Ponge (1899–1988), the *Mythologies* by Roland Barthes (1915–1980), the daily *billet* Robert Escarpit (1918–2000) published in *Le Monde*, the chronicles by Alexandre Vialatte (1901–1971), ... I am fond of the challenge brevity poses. This fondness goes back to childhood and becoming acquainted in elementary school with physical reality through the *leçon de choses*—to which I have devoted an entire book (1995), with that title.

My taste for short pieces also accounts for my entries into dictionaries and encyclopedias over the years. In the Nineties, the philosopher and Renaissance man Michel Serres (1930–2019) sought me out. Under his aegis, a small team— about 10-strong—of scientists wrote a thesaurus of the contemporary sciences, *Le Trésor. Dictionnaire des sciences* (Flammarion, 1998). We were at it for several years, starting in 1993; mostly during weekly stays at Fondation des Treilles, near Tourtour in the Provence. I cherish the recollection of our discussions, immeasurably wider and deeper than the finished product could be. The friendship I thus gained with Michel was a treasure as well.

About the same time I was involved with *le Trésor*, I was commissioned to contribute entries to dictionaries and encyclopedias published by Larousse, in Paris. Most exhilarating was my work on the *Petit Larousse*, the dictionary every French household possesses. I discovered that its entries on chemistry and chemists were both archaic and very poor in their coverage. I was given *carte blanche*, it was a joy to bring that dictionary up-to-date.

The years 2007–2013 saw me engaged in a similar task for *Encyclopedia Universalis*, a French equivalent and rival to *E. Britannica*. I did it shortly before the downfall of these high quality reference works due to the takeover of the Internet and Wikipedia.

In closing this digression about my taste as a writer for concision and short pieces, a word about my portraits in *La Jaune et la Rouge*. This a monthly magazine, of long standing, for the alumni of *École polytechnique*—where I taught from 1986 to 1999. In 2014, I volunteered to write each month the portrait of a living graduate from that *grande école*. This year (2021), a collection of these portraits, entitled *Figures polytechniciennes*, should appear. Each such portrait—influenced perhaps by Lytton Strachey's (1880–1932)—is only a couple of pages. They aim at showing the impressive diversity of lives and careers of the people featured.

About writing technique, I learned a lot during the years (1966–1970) I lived and taught in the United States. A major acquisition was invaluable in sharpening my pen: listing the points that had to be made, in no particular order, prior to organizing them logically or chronologically. My earlier French education had taught me instead, argumentative composition in three parts: you state a proposition and follow it by the pros, the cons and the conclude with a synthesis. The American way was much easier. In addition, I later found a book by the Franco-American Jacques Barzun (1907–2012), *On Writing, Editing and Publishing* (1986), to be extremely helpful.

My French education directed me to textuality, as opposed to orality. Upon first reaching the US at age 24, I was deeply impressed by the verbal skill—virtuosity of some Americans. By then however it was too late for me. I was and remained a person of the written, not the spoken word.

I do enjoy writing. Accordingly, writing results for publication in English, French being my maternal tongue, —primary publication—was no big deal. Here is a testimony (2020) by a former coworker, Arthur S. Greenberg (b 1946),

> I was always amazed by his writing skill- considering that it is his second language. The first article I co-authored with him (published 1970- almost exactly a half-century ago) was submitted to *Tetrahedron Letters* which then had a tightly-formatted 4-page form for direct photo-reproduction. I sat with Pierre as we (more he) composed and typed directly into the form and filled the four pages on the first try. No word processors. I'm not sure whether he even used white-out!

Since I found writing easy, I strove not to delay publishing our work, in order not to hamper the careers of my coworkers. It translated into an average of about five primary publications per year, during the nearly 40 years when I headed a laboratory. Review articles did not lag far behind: about one a year.

The impact of those articles, conjoined, as measured by the *h*-index and according to Google Scholar is 51: which means that I have published at least 51 papers having each been cited at least 51 times.

Jorge E. Hirsch (b 1953), the author of this index, suggested that, for physicists, a value for *h* of 45 or higher could mean membership in the United States National Academy of Sciences. Which is flattering—even though I do not belong to the group of the 2000 most-cited chemists. However, Geoffrey Bodenhausen (b 1951) and Antoinette Molinié (b 1950) published in 2010 a fierce critique of the abuse of scientometrics in general and of citations counts in particular. I could not agree more with their well-argued attack on narcissism, a rampant and devastating pandemic among us scientists, especially since the beginning of this century.

In my opinion, two qualities either mark or dismiss a scientist's worth. Expertise is one. It is overvalued in our era of intense specialization. It is evaluated through the *h*-index. That number, however, if it measures *depth* does not take *range* into account.

If I pride myself indeed on being serious in my work and, while being happy in its score, I put my pride in having followed my curiosity wherever it led me. I feel happy in having explored a few areas within chemistry—determination of molecular structure by physical methods; solvent effects; preparative chemistry using reagents on mineral surfaces; Monte Carlo numerical optimizations; interaction of ions with biomolecules; orientational dynamics of spherical molecules in the liquid state; etc. I have no regrets for having privileged curiosity over expertise, for having been a generalist rather than a specialist, during a time when the latter were leading the pack.

Whom does one write for in a primary publication? Predominantly, for a restricted group of peers, a few dozen people strong. Those are likely to have already heard the essentials, from talks given as seminars or at conferences. The written version is essential though, for the details in both the experimental part and the interpretation. Reception of our results was thus immediate, during my whole career, within the restricted group of peers. Such invariance, I wish, still obtains.

Which were the journals in which I published our research and why? The number one journal throughout my career was *JACS*, the *Journal of the American Chemical Society*. Its lofty position reflected dominance of the field by American scientists, or rather scientists from leading American universities, the Ivy League in the East, California in the West and places such as Chicago, Madison, Cleveland or Pittsburgh, Urbana-Champaign too in the Midwest. It went along with English being then the unchallenged language of science; with the post-Sputnik pouring of Federal monies into support of science and the rise of the research university during the Fifties and Sixties. The most coveted target for myself and my peers was at the time appearance of a communication (a "letter to the Editor") in *JACS*. I published 22 papers in *JACS*, of which seven were communications.

Notwithstanding my above claim of a wide range of interests, I was labeled a physical organic chemist. When I started chemical research in the Sixties physical organic chemistry dominated the subdisciplines of chemistry, it was highly regarded: for its ambition and rigor. However, the classical-nonclassical controversy about the most accurate description of the 2-norbornyl cation was suicidal, and synthetic organic chemistry toppled physical organic from its perch. I remained proud of my having been nurtured in it though, and named *laboratoire de chimie organique physique* my lab in Liège.

Indeed, in addition to *JACS*, the journals of the American Chemical Society I mostly published in—over 30 years—were the *Journal of Organic Chemistry* (*JOC*) and the *Journal of Physical Chemistry*: 15 and 3 articles, respectively. They dealt with preparative methods, on one hand, and electrostatic interactions, on the other. In *JOC*, the most successful paper was the very last, with Lionel Delaude as the co-author. He deserves full credit, with 235/399 citations (ACS/Google Scholar).

Not that I was not attracted to the physics side. Thus, I published primary results also in journals of physics, such as the *Journal of Chemical Physics* and *Molecular Physics*. I vividly recall John M. Deutch's (b. 1938) excitement when Erwin L. Hahn (1921–2016) at UCBerkeley quoted in an article of his, in 1966 or 1967, a joint paper of mine (1964), with Jeremy I. Musher (1935–1974) as co-author.

Which brings-up co-authorship. It has undergone tsunami-like change between then and now. Then: when I started as a scientist in the Sixties, the typical article bore two names, that of the graduate student, or postdoc, and that of the supervisor. Nowadays a typical publication carries half-a-dozen names.

My career-long policy was to not only list names alphabetically, also to grant authorship to persons able to present the paper in its entirety. Other contributors were acknowledged at the end, in the so-called Acknowledgment section. Such admittedly conservative habits have now been jettisoned. Any contribution, however small, has become entitling; such inflation I find unwarranted and worrisome.

But back to *JACS* as the leading journal in which one strove to publish. The first chief editor I dealt with was Marshall D. Gates, Jr. (1915–2003), of the University of Rochester. He was an Old-World gentleman, urbane and courteous, a pleasure to interact with. Martin Stiles (1928–2011), his successor, had a relatively short tenure (1970–1974). He was succeeded by Cheves Walling (1916–2007), who had moved from Columbia to the University of Utah.

At Utah, Walling appointed his younger colleague, my friend and co-author of the *Organic Spectroscopy* book, Peter Stang (b 1941) as one of his associate editors. This prepared Peter to assume the full editorship, which he did in 2003 until the end of 2020. Peter's tenure was one of the longest in *JACS* history, although not the longest since Arthur B. Lamb (1880–1952) served as chief editor from 1918 until 1949. In-between Walling and Stang, Allen J. Bard (b. 1933) of the University of Texas, a remarkable scientist, was an outstanding chief editor of *JACS*, from 1982 to 2001. I found him to be comparably kind and courteous as Marshall Gates.

Peter's tenure was marked by his rivalry with another Peter, Peter Gölitz (b. 1952), another great and pioneering journal editor. Gölitz was the long-standing (1980–2017) editor of *Angewandte Chemie*, published by Verlag Chemie in Germany, with two editions, in German and in English (*Angewandte Chemie International Edition in English* or *ACIE*). Both Peters shared the ambition to make their journal the best in the world.

ACIE carried review articles, which gave it an advantage over *JACS* in its impact over readers. Both journals ran neck-and-neck in their joint leadership in that respect.

I published reviews not only in *ACIE*, also in *Accounts of Chemical Research*, a review journal of the American Chemical Society (that published *JACS*).

To publish was to benefit from peer-reviewing. In being a referee, I must have received two or three dozen manuscripts yearly to referee. The remarks by referees were often perceptive, if not profound; they helped to improve one's papers. I have a vivid recollection of the referee of a manuscript I submitted to *JACS*: he went so far as to redo my experiments in his laboratory in order to check them. This was Robert J. Kurland, of the State University of New York in Buffalo. I was very grateful to him for being so conscientious.

I found it congenial to write review articles on our work. They obeyed an implicit rule of the genre, to be cumulative and sum-up a number of primary publications. Rather than being self-questioning or even self-deprecative—colleagues see to that. My main regret in that vein is not having let more of an exploratory urge shine through. Prophecies in general turn out to be wrong-headed, I was keenly aware of that. In any case, judging by the number of citations they gathered (between 400 and 500 to date), my most successful review articles dealt with chemical reactions on clays. They date to the end of the 1980s and were published in *Accounts of Chemical Research* and in *Science*.

Being an editor myself: I mentioned already my stewardship of the *Bulletin* of the Belgian chemical societies—there were two such societies, one for Flemish speakers and for French-speakers. It taught me the job of an editor. Not only lining-up at least three referees competent to review a submitted paper, making sure they would comply in timely manner, getting the senior author to answer carefully the reviewers comments and, last but not least, improving the language of the article—its title, the vocabulary and syntax, the spelling, the organization, ...

I did not dislike the job but I did not care for it particularly. I did it out of a sense of duty. To whom? To the scientific community worldwide and out of self-respect. In later years, during the 1990s predominantly, other editorial responsibilities came my way. Wallace S. Brey (b. 1922), the founder of that periodical, invited me to the editorial board of the *Journal of Magnetic Resonance*, the most highly regarded journal in that area. I took this appointment as a major honor. Only light work came my way from this responsibility.

Then, Joe Lambert (b 1940), still then at Northwestern University, whom I regard very highly and who is a friend, appointed me as a regional editor (Europe) for the *Journal of Physical Organic Chemistry*. My wife Valerie served as a volunteer editorial assistant, her help was indispensable, it was a lot of work and many manuscripts were submitted to that journal (Fig. 12.2).

I have two main recollections from that editorship. This was a transition time for science as a whole. The Third World was coming in. I saw it as my responsibility, not only to uphold high standards, also to help in brotherly manner those newer colleagues to get into print. Quite a challenge: their mastery of the English language ranked between barely adequate and awful. I saw it as my responsibility to gently improve their write-ups in order to make them publishable.

My second recollection has to do with contributions from former Warsaw Pact countries after the collapse of the Soviet Union in 1989 released them. I have a

Fig. 12.2 Valerie hiking in
the French countryside, 1996

distinct recollection of a Polish colleague, who was doing passable work but nothing original or distinctive, swamping us with manuscripts.

About 1990, I turned down a position of power. Guy Ourisson (1926–2006) asked me to succeed him as the European regional editor of *Tetrahedron Letters*, then the leading journal for fast communication in organic chemistry. I declined this offer, I was indeed too busy and spread out and suggested instead that Léon Ghosez (b. 1934) be chosen. He jumped at the opportunity.

Being an editor also included collective books: lining-up fellow authors, pestering them to write their chapters and turning the whole into an attractive, if possible comprehensive and readable whole. I accepted a few such tasks and they turned out as useful contributions to the chemistry literature.

I gave numerous talks on our work. During the 1980s, I had a major incentive for doing so. When my eldest daughter, Sophie, turned 16, she completed her high school studies in Belgium. But where could she go for a higher education? A thorough investigation on my part showed that European countries—Belgium, France, Germany and the UK, in the first place—would all impose an early choice of a career. Only in the United States could she get a general education without enforced early specialization.

In the end a liberal arts institution became our choice, Smith College in Northampton, Massachusetts. It ranked very highly for the quality of its education and Sophie was accepted there.

However, since we lived abroad, Smith demanded payment in full for its tuition for the four undergraduate years. While I do not recall the exact amount, it was about US$ 40,000 per year. I was footing that expense by myself, with no other help. My salary as a professor at the University of Liège was comfortable but it was clearly insufficient to cover that big expenditure.

Accordingly, I turned to traveling frequently to the US and to presenting seminars there in a variety of universities. At the time, you would be offered an honorarium for each such talk. This bounty enabled me for those 4 years (1981–1985) to foot the Smith College bill. In the process, I visited many small university towns. I must have been an acceptable lecturer. Hence, I had little trouble getting invitations to give a seminar.

At the beginning of this century, after retiring from my permanent, European appointments, in Belgium at the University of Liège and in France at the *École polytechnique*, I visited Berlin a few times. I was invited at the Max Planck Institute for the History of Science (MPI). Later on, I attended conferences on the history of chemistry, also organized at the MPI by Ursula Klein (b 1952). These activities coincided with my eldest child, Sophie, who had become in the meantime a diplomat, living in Berlin with her family and working at the French Embassy.

During one of these stays, I visited the Berlin office of the Springer Verlag publisher. There, I met with the chemistry editor, Dr. Marion Hertel. She divided her time between Springer offices in Heidelberg and Berlin. We agreed that I would write a book of my various experiences as a scientist, regarding publication in the widest sense.

Which I did. The book, *Communicating Science*, was published in 2007, in a most attractive format. My expectation for younger scientists to take it as a guide and treasure it as a companion was frustrated. In terms of sales and impact, it was a flop. The reasons why I guess, are multiple: insufficient promotion by Springer; targeted people too busy and convinced they did not need any such advice; perhaps also, put off by the style, light and humorous whereas they perhaps yearned for a more solemn approach. Who knows?

I can only blame myself for the paltry influence of that book. Perhaps I should have gone around pushing it through seminars and workshops.

However, self-promotion is not my forte. Teaching in the United States made me intensely aware of the need for self-promotion. Some of my co-authors and colleagues, Jerry Musher (1936–1974), Paul Schleyer (1930–2014) and Carl Djerassi (1923–2015), in succession and to mention only these three, were outstanding at it. Why did I remain reticent about this very American form of salesmanship? As Paul Schleyer once told me, "if you do not refer to your own publications, who else will?" It was not for lack of self-confidence. Nor a feeling of inadequacy. Deep down a natural shyness. A definite taste for understatement, more British than American. And an abhorrence for anything smacking of commercialism, ingrained in me by my family and my whole entourage during my formative years.

To return to *Communicating Science*, there was one instance of a resounding success, for which it deserves and received credit. I had been approached in 2008 or 2009 by Jean-Philippe Échard (b 1975), a young chemist whom I first met in 2004 or 2005, when he was serving an internship with Dr. Barbara H. Berrie (b 1955) in the laboratories of the National Gallery of Art in Washington, DC.

He had had an article of his, together with Loïc Bertrand and other coworkers, rejected. Their paper dealt with the composition of varnishes on Stradivari violins. I rewrote it and suggested that it be resubmitted to *Angewandte Chemie* (*ACIE*). It was then promptly accepted. The cover of the issue in which it was published (December 2009) even featured a photograph from that paper. At the same time, I directed Jean-Philippe and Loïc to my *Communicating Science* book—which they raved about. Their *ACIE* paper, to his day, was cited 70 times—a success.

This chapter on science writing would not be complete without mention of a negative, viz. what I could have written but did not. Exhibit Number 1 in this respect is a textbook of chemistry, aimed at first-year university students. I came very close to publishing one, based upon my teaching at Cornell in 2001 and 2002. It would have been entitled *The Wonderful World of Chemistry*. I even had a publisher lined-up, Wiley in New York. The pros were heaps of money I could have earned from such a venture, which can be hugely profitable. The cons were embracing slavery: for the following 10–15 years I would be tied to that book, to the demands of teachers using it, to fighting competition from other books in that niche, constantly issuing newer versions and, in general, remaining on top of that game. I am not unhappy about having renounced this particular costume, however tempting it was to start wearing it.

I could draw for my decision against on a precedent. I had published, in 1995, also with Wiley, a textbook of organic chemistry based on my classes at *École polytechnique* and entitled *Organic Reactions, Simplicity and Logic*. It had small sales, it was a *succès d'estime* as one say in French.

While I did publish a textbook of quality, I met with better commercial results with science popularizations, in French as a rule—even though some received a number of translations. Two of the English titles are *Salt, Grain of Life* and *Citrus*, published in the US with Columbia University Press (2001) and The University of Chicago Press (2007), respectively. I loved writing such essays, with Lewis Thomas (1913–1993) and Stephen Jay Gould (1941–2002) as models. They met with good success, amounting for about 10 years to an additional thirteenth month in income. Paperback versions were issued. In my native country, in France, they earned me significant prizes from the *Académie des sciences* and *Fondation de France*. I chose to enter that avenue primarily from a sense of duty, a moral imperative to inform my fellow-citizens.

In like manner to a trigger-happy soldier, was I an irresponsible writer in all kinds of formats and just a polygraph? My hope is for the preceding pages to give a negative answer to this rhetorical question. Every single piece I wrote was, true due to a facility with pen or keyboard, it was moved also by an overwhelming curiosity for many a nook and cranny within chemistry. Thus, a polymath of sorts, not a polygraph.

I'll turn to my beloved eighteenth-century for justification. In England, Joseph Priestley (1733–1804), in France Denis Diderot (1713–1784) wrote earnestly on very many subjects. History has given them their well-deserved pedestals. Far from comparing in quality my output to theirs, I hope that my chemical writings will be judged on their own merits—as an attempt at communicating my sense of wonder for the material transformations nature has graced us with.

Teaching = To Raise

13

This chapter will deal with my teaching experiences, in Europe and North America; with my fondness for the job and, even more important, discovering new resources and improved manners for conveying the material; as implied by the title for this chapter, teaching is akin to raising children, i.e., in this case, bringing the students to an adult and responsible level in science.

My starting point is play. There is a game for young children in the United States and other English-speaking countries known as Show and Tell. A common classroom activity in early elementary school, the pupil is called upon to show an object in the classroom and tell a story about it. The etymology of the verb "to teach," in English as in other languages, Greek, Latin, French, Portuguese, what have you, points likewise "to show with words, to present in speech, to tell the meaning of some sign, etc."

The previous chapter was about the production of textual reports. This one deals with the power of spoken words in communication, by what was known in the Ancient World as "oratory." I mentioned earlier my internship in a Parisian *lycée* with Madame Barberon and her emphasis on vocal skills, especially on learning how to set one's voice to talk at length with no fatigue and a volume comfortable for one's listeners. In so doing, I discovered a booming voice of mine—which could become useful in some circumstances.

Another early experience came from teaching graduate classes at Princeton: in preparing for a class, it gave me discipline to go for the original material rather than relying on secondary literature. This is the same as the axiom in textbooks published by Bill Benjamin, viz. why not get world-class science leaders address directly first-year students? Or with the equivalent guiding principle at *Collège de France*, in Paris: professors there are expected to lecture about their ongoing research, whatever their field. In other words, a Show and Tell attitude, but from the frontlines of science.

I did emphasize so far the teacher as lecturer. However, there is another aspect, equally important, in the make-up of a good, a memorable teacher: student participation. I learned it already in Paris, in *classe préparatoire*. If you want nowadays to witness a superb teacher at work, go to Lycée Louis-le-Grand and attend a class in mathematics by Nicolas Tosel (b 1967). Or, attend a class in linguistics by John Goldsmith (b 1951) at the University of Chicago. You will witness from both these teachers *collective* elaboration of a lecture by everybody in the classroom, teacher and students; that demands preparation by the teacher, a deeper, more intense in scale, than just a standard lecture.

Was I lucky! My teaching career started at Princeton University (1966–1970). It was graced there with gifted students and colleagues, some of whom (Hubert N. Alyea (1903–1996), Edward C. (Ted) Taylor (1923–2017), Paul Schleyer (1930–2014), Kurt Mislow (1923–2017)) were outstanding educators. I learned a lot in my attempt at emulating them.

My career proceeded with a full professorship, aged only 31, at the University of Liège, in French-speaking Belgium: it was eerie, a step back into the nineteenth century, with autocratic professors—many mediocre—in the German style. I had to become a reformer there, I had to innovate. My Liège appointment as an *ordinarius* covered the period 1970–1986.

I introduced multiple-choice question tests—a tool I had learned from the Educational Testing Service in Princeton. Other innovations of mine were faculty evaluation questionnaires; replacing an all-determining final exam with continuous testing; replacing mimeographed lecture notes with published textbooks; showing movies in the classroom; etc.

To further hone my skills, I collaborated with the world-renowned expert in pedagogy, Gilbert de Landsheere (1921–2001), a colleague in the humanities. I underwent for instance an autoscopy in front of a video camera. It showed me some of the unconscious distracting details in my body language which thus I was better able to control.

I received in 1986 a call to return to my country and become a professor at the *École polytechnique*, a leading institution, comparable to Caltech or MIT. I taught there until I took early retirement in 1999. During those years, I did not totally renounce my professorship in Liège, becoming an *extraordinarius* there in order to be granted in due time the pension I had already earned.

At the *École*, the caliber of the students reminded me of my Princeton days. Lectures were rather colossal undertakings, each was meant as a window into a whole chapter of chemistry. A rule for a university teacher is the 1:1, viz. an hour of preparation for an hour of teaching. At the École, it had to be turned into another, a 10:1 rule.

As I chose to take early retirement, aged 61, at both the *École* and the University of Liège, I was concerned about missing the thrill of teaching. However, there again, I was very lucky, receiving a call to Cornell as a visiting professor in both 2000 and 2001. In this manner, the transition from educator into active retirement as a writer was made considerably easier. At Cornell, an Ivy League institution with again first-class students, I was given a responsibility skirted by my colleagues in chemistry, to teach a course to initiate into science non-scientific students. It was a university-wide requirement that many of these students resented. That challenge, I found highly motivating. I shall mention it again in the remainder of this chapter.

My visiting professorship at Cornell was not the only one. In earlier years, I served visiting professorships at the universities of—in alphabetical order—Berkeley, Chicago, Colorado, Connecticut, Hamburg, Kansas, Lausanne and Toulouse.

I became a teacher out of a sense of moral duty. That of sharing my knowledge with fellow-citizens. I had been lucky with having become a scientist. I thought it was the best job in the world. Hence an obligation to share my knowledge with others. Not so much my knowledge of chemistry as my constant wonder at the beauty of the world. Not the specialized toolbox I used for acquiring scientific knowledge, it was too intricate and any mastery was too time-consuming, but the way in which to make sense of the results it provided. I tried to give my students some critical acumen, providing them with simple guidelines for evaluating all sorts of things, such as sports results, fluctuations of the stock-market or other diverse data.

For instance, I addressed with them the question of evaluating the prediction of how 50 million people will vote based on polling a thousand persons. What does one mean by air or water quality? I relayed to them the gist of some major discoveries, for instance elaboration of chemotherapy by Paul Ehrlich (1854–1915). I tried to educate them as consumers: for instance, on the differences between compositions of gasoline, kerosene, diesel fuel and paraffine wax. Thus, I showed them science as a web of interconnected concepts.

What else did I bring into my teaching, or not, and why? I'll start with the negative: no, or very few, classroom demonstrations. Students like to watch them, not for learning sake but out of a kind of voyeurism: in chemistry lectures, spectacular shows, explosions especially!

Conversely, I showed movies. At the University of Liège, my lectures—to a big audience of about 200 students—independent of my will were scheduled to last an hour-and-a half. Few adolescents can maintain their attention that long. Hence, I took advantage of our proximity to Brussels. The Free University there had a collection of science movies from which I was able to borrow—usually films lasting 10–20 min that suited nicely my purpose.

Which brings up technologies for displaying information and their changes over time. During my career, from the 60s into the late 90s, these changed from the blackboard to screens of various types, adequate to project slides, transparencies, and finally, computer-generated images. The worst such innovation was having to use an epidiascope—a projector of transparencies—it freezes the lecturer in a crouching position over the machine, which I disliked: moving about, in my opinion, is essential for the teacher to hold his/her audience attentive. Each such medium suffered rapid change. Slides went from black-and-white, and Polaroid; to white text on a blue background; and to color, at long last. Such a relief, when instead of photography computers enable us to prepare our own slides.

My blackboard technique was adequate. I learned quickly not to mask with my body what I had written and I developed the skill for not wiping out the board too soon for the students—whatever you say, some are intent upon noting it. Of course, my generation of chemists had the example of R. B. Woodward (1917–1979) to try and emulate, he was a true artist of lecturing at the blackboard.

I liked slides and showed them, about one a minute during a 1-h (50 min) lecture. Each slide had an image and some text. I strove to make each intriguing and not explicit without the accompanying spoken comments.

I experimented mastery learning. In the so-called Keller Plan, the entire subject matter is first divided into chapters—a dozen, for instance. A student is allowed into the next module only after he or she has demonstrated satisfactory mastery of the previous. It works; but I found it to be highly time-consuming. Teaching becomes a set of individualized tutorials. Thus, I tried it for a single term only.

As their originator and presenter, I am unable to convey the flavor of my lectures. But I feel allowed to depict their intent and describe their make-up. A guiding principle was relating to the students, addressing their concerns and interests. I tried to always speak in simple terms. Likewise, I used carefully chosen or drawn graphics to give explanations and keep them as simple as possible. As a courtesy to the captive audience, I made a point of showing the progression in the lecture as it moved along; and to structure the presentation in a small number of parts, usually 3–5. And I spelled out the take-home lessons at the end of each lecture.

I took care in explaining the importance of each topic I chose to present. I made my choice of examples wide and informative. Indeed, I tried being as informative as possible. I often mentioned current events as shown in television and newspaper news. I would refer to science in the making, to current reports of advances and discoveries. I would also accurately refer the statements I presented.

And tried to be funny whenever possible.

An important issue, I found out, was to bolster self-confidence among one's students; freshmen especially. They had heard of chemistry as an arcane subject. "You just have to memorize it, there is nothing to understand" is all too often their mantra. How to counter it? Thus, I saw it as my task to start my classes by having them gain self-confidence.

My way of doing it was by having them answer order-of-magnitude questions. Anything will do, it does not have to be chemical. Examples, from the back of my head: what is the order of magnitude of our distance to the Moon? the size of Australia? of Jamaica? of the aspirin molecule? the volume of the lung? the weight of a human kidney? the body weight of a gorilla? of a bonobo monkey? the volume of fuel embarked in a liner jet, a Boeing 777 say? the weight of food in your refrigerator? in the refrigerators of a large cruise ship? the weight of a standard pee? of a pencil? Etc. etc.

By telling them the accurate answer and showing that their guesses had come close, it would allow students to realize the impressive amount of knowledge their minds had already registered; and their ability at guessing more or less correctly the answers to a host of other, just as unlikely, questions.

These were oral exchanges, between me and the students. I deem it important for any teacher to engage in dialog with the students. This of course is easier in a small classroom. But one should not renounce this attempt in a large classroom of 200 or 300 students. There are ploys to succeed in such an admittedly ambitious, but absolutely necessary aim. Since I was able to succeed at that, I am confident that colleagues will have the imagination and the guts to do it too. The rewards are there: once you have struck such a rapport with your students, they will trust you as their guide and follow you—anywhere.

Any good, self-respecting teacher should be able to convince his class of the relevance of his subject to both their life and their chosen career. It came easily to me, with a caveat though: as a chemist, I had to address minds prejudiced against the subject—especially when it was imposed on them by some university regulation, rather than it being their choice. I encountered it with Chem 106 at Cornell, at the beginning of the century. This was a course of science initiation, it was compulsory for all undergraduates not having a science major. A good idea, but not very easy to implement!

My approach was to build upon a statement, "Chemistry, the central science", that the American Chemical Society has made into a slogan. But I'd better cite some examples.

I gave a whole lecture in 2001 or 2002 on what I termed "the 2003 Catastrophe." My scenario had a meteorite the size of a football field hit the area of Baton Rouge in Louisiana, where much of the American chemical industry is located. I described some of the consequences, not only to the American economy, to the global one of that devastation. Eerily, when Hurricane Katrina hit that very area in 2005—New Orleans and its region—the destruction resembled my predictions.

How to interest biologists in chemistry? I would use chemical communication and warfare as my leverage tool. I would refer to the feud between plants and herbivores: how plants warn one another of an incoming attack and how they defend themselves with chemicals toxic to their predators.

Archeologists? This was easy, because I had served an internship in Marsal, in Eastern France, helping to dig up a proto-historical array of kilns serving to boil brine for salt manufacture. I was able to tell, for instance, of my having analyzed the raw material—too rich in strontium, which would have caused chronic illness among the workers (Fig. 13.1).

Green-minded students of nowadays: this was easy, I would tell them stories of chemists—Svante Arrhenius (1859–1927) first, but also his successors Sherwood Rowland (1927–2012) and Mario Molina (1943–2020) who had announced and warned about the severe consequences of global warming.

All of this serves to show that, not only in research, in knowledge-sharing as well, my much-encompassing curiosity helped very greatly.

Even a cursory look at standard textbooks of chemistry shows a most disquieting growth in their size during recent decades. To demand mastery by the students of such an amount of material is unrealistic. Moreover, many students have no particular inclination for the subject. In the United States, for instance, premeds have been required to take organic chemistry for many years, for no good reason except for the memorizing effort. One might as well assign them rote learning of a telephone book!

I published some years ago (2013) an article on this topic. The solution I recommended is a radical change of approach. I advocated teaching chemistry as if it was a foreign language!

Fig. 13.1 A segment of the dig in Marsal (August 2005). Salt-making kilns (salmon colored) are juxtaposed to circular brine tanks lined with blue clay. Their high number on the site points to mass production, dating back as early as the seventh century BCE. Numerous accompanying artefacts, from the Mediterranean in particular, point to long-distance trade (photograph by Bruno Wirtz)

I based this recommendation on chemicals being assemblies of atoms and groups of atoms—just as verbal utterances are made up of phonemes. Hence, my solution to an infuriating escalation. I should have tackled this particular problem many years earlier and I ought to have started to devise explicitly a solution—not only pointed the direction to it. I hope that somebody in due time will take it up again and address decisively the swelling of chemical knowledge before it drowns chemistry as an academic discipline.

What I liked best about being a teacher? Interacting with many people. Since by nature I am shy, my position of power—to be honest about it—made it easy, seemingly natural. Not only the interaction, its one-directionality. People attentive to what I said. I loved the freedom too, the freedom to roam about on any subject I chose to tackle.

I liked the discipline too, needing to deliver on set times when my classes met. I never missed one in 40 years of career. You might be sick, suffer from a cold, an indigestion or a headache—nevertheless, you had to get up, go out and meet your class.

The worst happened in 1982, after I had been felled by a virus. Known as poly-radiculo-neuritis, it is a viral infection inside the spine. This condition is a close relative to polio and to the Guillain-Barré syndrome. Muscles abandon you. Weakness and fatigue overwhelm. Nevertheless, I did not miss a class.

In 1996, while suffering from painful sciatica from a bad back, I even taught on my back, supine on the desk. I had gone on a trip to Nouméa, New Caledonia. My wife Valerie and I spent a week there. I lectured for a couple of hours daily. How? Lying on my back, of course.

What I liked least about teaching? Administering exams, for sure. I did conscientiously that part of my task, but did not relish it. The reason for my unease were apprehension from the examinees, for one. I also did not care for the paperwork, the reports one had to submit to the administration—namely colleagues who, for various motives, had opted out of the most gorgeous job in the world. I was of a mixed mind about the adulation from some of the students, both relishing and hating it. As yet other reasons for disliking my chosen job, there was the time it took away from doing active science rather than just being a passive mouthpiece. And the notion of perhaps not having been capable of an active role in life: as G.B Shaw put is so memorably: "those who can, do, those who can't, teach."

The characteristics of a good teacher? Witty. Self-deprecating. Quality of the speech delivery, slow and clear. Initiating a great deal of interaction with the audience. Setting the highest standards for himself while being tolerant of the numerous inadequacies among the students. And a precious form of respect: it has always struck me that the whole is, mysteriously, more than the sum of its parts. A group of people listening to you is worth your utmost consideration.

Before I started lecturing, I would make sure of my breathing, even went outside and took several deep breaths, filling and emptying my lungs—thus oxygenating the brain. I would then appropriate the classroom to myself, i.e., walk back and forth in silence while students watched for a few instants. This served to start to focus their attention on the person about to address them.

The first words uttered are important, they define the tone and tenor of the lecture. I found it a good idea, from my earliest times as an academic, to write out in longhand my opening paragraph (the *incipit*, to use the appropriate term)—making sure that it be elegant and simple, easy to say and memorable. Indeed, I would memorize it and deliver it carefully, in order to open the lecture with modesty, gracefulness and humor.

What were the rewards of being a teacher? They were numerous. I'll mention them in no particular order. Being my own master, with no one to dictate the subject matter or what approach to take. Public consideration for undertaking such a task as educating the young—and, together with it, lack of public consideration for accepting such a low-paying job! Appreciation from students—better yet having some become your friends. Also, there is nothing an educator likes best than having former students come and say hello.

Why did I become a teacher? Imitation; because I was graced with admirable teachers, most of whom this book has already mentioned by name. Wish-fulfilment, not mine, my father's. An engineer, after taking his French degree in the middle of his adult life (France did not recognize his Hungarian credentials from the Technical University in Budapest), he longed to become himself an academic—but did not.

Another reason is my training: most of us, having attended the NSE *classe préparatoire* in Lycée Saint-Louis in Paris, during the second half of the Fifties, have indeed become university professors, teaching various sciences: physics, biology, geography, geology, predominantly. Yet another factor in this vocation of mine is having been bookish since my early years and thus fascinated with knowledge and its transmission. Educate means bringing up and I love the notion of talking to a person and, in this manner, quench his/her thirst for knowledge. Are these enough factors?

A key question is: whom are you addressing, whom do you talk to? In my teaching, I strove to maintain respect and to never talk down to my audience. Whatever its age and level, whatever their knowledge of the subject matter, I always assumed significant wisdom on their part. Whatever the size of the group, ten to twenty or several hundred, it would include, I trusted, worthy recipients who would benefit from what I was trying to impart. Such an uncertainty always accompanies teaching and makes it worthwhile as a dash into the unknown, not unlike scientific research.

Is it being narcissistic to enjoy being the center of attention for 50 min or so? Surely. And yet this is only a minor aspect. It does not explain why we teach and, at least as important, why students pay attention to what we say. The teacher-students interaction is clearly a form of parenting, with enforced listening, experience sharing, a dose of mutual affection and constant encouragement to do better. There is a dynamic of the classroom and it takes little experience to feel it: sudden attentiveness upon mention of the topic of today's lecture, its boost when everyone is striving to follow your explanations and the collective relief when you announce the end of the class.

A rather weird resemblance is inescapable, to a religious service.

It is a responsibility one carries as an academic. It is the part of the job I liked least. Part of the job? Not really. It is more a social function than one pertaining to transmission of knowledge. But I did it and I must have examined hundreds if not thousands of students in nearly 30 years of career.

Students, justifiably, fear and even dislike exams. As a student myself, I did not have too difficult a time. A couple of anecdotes first, to show that far from being unusual I was rather typical—cheating included, I am a little embarrassed to confess.

During the written part of the competitive entrance exam (*concours*) at the *École normale supérieure*, I was seated close enough to my friend Alain Renaudie (1937?–1960) we started in hushed tones of course, to discuss the test. His sister Josette Renaudie (1930–2018), older than he and a mathematician, already a member of the École, stood in the room, together with a few other adults, to ensure that nothing untoward happened. She rushed to our side and forcefully hushed us up.

In another entrance exam, to ENSET (*École normale supérieure de l'enseignement technique*)—one in fact I had no genuine wish to enter, my desk was next to that occupied by a young lady there for a different exam. We just traded papers, I wrote her essay in philosophy and she gave me the solution to the problem in maths that I was assigned.

A couple of years afterwards, while studying at the Sorbonne for my BS, I was taking an exam in physics. The neighboring desk was occupied by a nun—this was prior to the Vatican 2 Council, and she wore her ecclesiastic garment. During the whole procedure, that lasted 3 h I believe, she tried hard to read what I was writing: should I be ashamed or proud of having prevented her from a serious sin?

One of the intermediate stages between being a student and becoming a professor is serving a teaching assistantship. This was in Paris where I had such a function in biochemistry at the Sorbonne with Professor Lederer.

I recall this revealing incident: I was supervising laboratory work. For safety reasons, extraction with organic solvents, volatile, toxic and flammable, had to be done outside, in the courtyard. A young lady-student who looked bemused that the liquid would not flow out of the glass container once she opened the tap at the bottom—she had not realized she had to open also the top of the apparatus. I exclaimed, "*mais enfin, Mademoiselle, réfléchissez un peu*" (why don't you give it some thought). To which she retorted, "*mais je ne suis pas là pour ça, Monsieur*" (but this is not why I am here, Sir). This answer alerted me to a rift between students and teachers, getting a degree and getting an education are not necessarily synonymous.

Having become an academic, my first experiences in administering rather than taking exams were at Princeton University at the end of the Sixties. Exams at Princeton were run under an Honor System: students were trusted not to cheat, as the gentlemen (and ladies, as the university became coeducational) they were nurtured into. Accordingly, examinations were not supervised. However, if a student was caught cheating, at least in principle it brought about immediate exclusion from the University.

I recall mostly the written examinations graduate students had to take on a monthly basis. These were the cums, short for cumulative examinations. The rule, if I recall it properly, was that they had to pass eight such exams by the end of their third year.

And these were tough exams. Each month four or five faculty members would put together a test, comprising questions in the various sub-fields: organic, inorganic, physical and theoretical, polymer and analytical chemistry. Students would try to guess which faculty members were devising the test, to try and best prepare themselves for the perceived personal equations of given professors

If cums were tough, the final exam—that for the granting of the Ph. D. degree—was the toughest. At that time, the end of the Sixties, the system had two phases. The written dissertation had to be approved, not only by the supervisor, also by another faculty member. Once this hurdle had been cleared, the student had to present and defend publicly, in front of a few faculty members, no fewer than ten different research proposals. I believe this rule was borrowed from Dutch universities. It was too strict and, at the beginning of the Seventies, the Chemistry Department watered it down, bringing down the number of proposals down from ten to three, in a first step, and later on, from three to only one.

Following my years at Princeton, I moved to Belgium and I found there, also for the doctoral final examination, yet another instance of influence from the Dutch usage. This was at the ULB (*Université libre de Bruxelles*). The research accomplished, already written-up in a dissertation, had to be presented and defended twice, first in a *défense privée* and only afterwards in a public defense. The former was serious and elaborate. The candidate faced and answered tough questions from board members. It resembled a research seminar, in that you had to convince the deliberately doubtful of the seriousness of the work, of the validity of its results and their interpretations. This lasted for several hours. Some time later, a few days typically, there would be the public defense, a formal occasion with friends and family present, for the new doctor to shine presenting his or her work. I recall in particular having served as an external examiner—a devil's advocate—in the final doctoral exam of Isabelle Stengers (b 1949), who would become a highly public European intellectual. During her public defense, I chatted delightfully wth her father, Jean Stengers (1922–2002), an historian of note.

When I arrived in Liège in 1970, I discovered an entirely different system of exams from Princeton. The student body, as a major difference from American Ivy League universities, did not consist of an elite selection. It rather represented the Walloon society as a whole. On paper, it was much more democratic and even egalitarian—which I found to be in many ways a Belgian insistence. In addition and by law, exams had to be administered orally—they were indeed a huge waste of time.

What they consisted of, in the vast majority of courses, was rote memorization of the typically mimeographed lecture notes. Very few among my colleagues demanded a genuine understanding of the subject matter. I have recounted elsewhere in the book my introduction, instead of testing at pre-ordained dates using multiple choice questions.

The population in the Liège region, also known as *Wallonie*, was then very much segregated by social class. Industrialization occurred in the nineteenth century, it was one of the earliest regions in Continental Europe to undergo a transfer of technologies from the UK. The dominant industry was steel metallurgy: the coal was from local mines and the iron ore came on a railroad from Eastern France. Labor consisted of a proletariat of imported workers, of three major origins, Italy, Poland and Flanders. Accordingly, the student body at the University reflected this make-up.

The Liège population was made of descendants of these workers and of the dominant, high bourgeoisie, scions of Belgian engineers—the figure of the engineer still had in the Seventies a high, almost mystical status —, and of other liberal professions (MDs, attorneys, notaries, judges, . . .) together with remnants of an aristocracy. This domineering social segment elected university studies as one of the passports into ruling positions. Exams at the University of Liège in the Seventies thus were noteworthy. They were a rite of passage, of the kind anthropologists have reported and studied. Students would dress up for it. It amounted to an initiation into the local (and ruling) bourgeoisie.

An anecdote will complement the point about bourgeoisie values overruling by far acquisition of knowledge. One evening, my wife and I went to Brussels. We had tickets for Ballet du XXe siècle, the company led by the choreographer Maurice Béjart, at Théâtre de la Monnaie. As we were walking to take our seats, a young man rushed to us and, almost kneeling, called me his savior, his benefactor. During the few minutes before the music started, he told me his story. He had been among my students at the University of Liège, enrolled in the first year of biological studies. At the end of the year, he failed all his exams—except for mine. Hence, he had to repeat that first year. On that second try, he passed all final exams—except for mine. And at the end of the oral examination, I had told him that he was clearly not made for scientific studies, his talents were elsewhere. Which gave him the courage to confront his family and tell them that dance was his true call. He had become press attaché at Ballet du XXe siècle. He added that pressure from his family towards studying the sciences came from his engineer father and his grandfather a professor of engineering at the University of Liège.

I stood out from a majority of my fellow-professors: I was perceived as a stern and difficult examiner, because I insisted on the student showing a genuine understanding of the material. I was called a *moffleur*, i.e., someone who failed students and gave them poor grades. Indeed, when the Europe-wide Erasmus student-exchange system came into existence (1987), some of the chemistry students at the University of Liège took advantage of it to circumvent my exams: I had the reputation of being way too demanding.

My insistence on in-depth understanding of the material was anathema to quite a few of the students. The unwritten contract, as they understood it, was to pay lip-service to their teachers and content themselves with regurgitation of what their classes had consisted of—to insist upon genuine understanding and showing ability to transfer some knowledge to a different context, as I did, made them not only unhappy and uncomprehending.

At that time of the Seventies and Eighties, in Belgian universities professors were required by law to administer oral final examinations. What did these oral exams of mine consist of? As I have made clear, I did not ask students to regurgitate some segments of my lectures they would have rote memorized. I was more demanding, I wanted them to demonstrate some understanding of the material. I devised exercises for that purpose. Typically, students were given half-an-hour to prepare their answer after having been handed the text of the test. I would then spend some time, about 20 min per student, for presentation of the answer; to find out what level of understanding had been achieved.

One year, towards the end of the Seventies-beginning of the Eighties, I realized these one-on-one interviews provided me also with a unique opportunity to probe the personality of an individual—not only to gauge answers to a chemical query. At first, this approach was successful. It brought to light facts and features that sometimes were highly disturbing.

I have thus a vivid recollection of a bright young female Moroccan. She was failing to solve the chemical exercise she had been assigned. Talking to her I realized that her studies were not her main task. In addition—or alternatively—she was taking care of the three brothers with whom she shared an apartment in Liège: She was cleaning and cooking for them, doing their laundry, ironing and so on. I was appalled that they had imported from North Africa their culture of feminine subservience to the males in the family. What could I do? Given her harsh circumstances, I believe I was indulgent.

But this type of examiner-examinee relationship showed no durability. I indulged in it only for one exam season or two: I realized quickly that students were passing to one another notification of my being personally inquisitive in this manner. Some students started showing up with personal problems even before I had even started probing. Hence, I had to regrettably drop that part of the interaction with the examinees.

The other institution where I was a professor was the *École polytechnique*. Its students were of the caliber of the Ivy League, they were as a rule very bright and apt at easily taking and passing exams. One of my recollections is of having given top grade to a young man. Before he left, we had a short chat: I asked him how he had prepared for this exam. He answered that on the evening before the last he had picked up the lecture notes and started reading them. They read easily, he added. I was impressed by his ability at mastering a wide body of material from investing a few hours—half-a-dozen, is my guess—just reading through.

The *École* had its version of an honor system. Students took tests from their own individual room, in a *casert* building—this was the name for student housing, short for *casernement*. Such indulgence came from imitation of Anglo-Saxon rules. Which serves to show that what obtains in a given culture may not translate into another, different culture: ungentlemanly behavior does not translate easily into French.

I have thus recollection of a test in biology. The professor who provided and graded it discovered that about half-a-dozen students had cheated—how they did it, I never found out. One was the son of a fellow-professor. Anyway, they all got a grade of zero and were punished. The year? About 1992.

Exams at the *École* and their grades had ranking the students as a function. The grands *Corps de l'État* made (and still make) their pick based on this ranking; starting with Corps des Mines which picked the ten best students and then trained them for heading the leading French businesses and public administrations. This was a key part of the French meritocracy. As two American historians, Carl E. Schorske (1915–2015) and Jerrold E. Seigel (b 1936), remarked to me, during a joint dinner in Paris in the 1980s, the French are obsessed with rankings, not only of wines and restaurants, of people as well.

Were there differences in exams between the start of my career in the Sixties and its ending at the turn of the century? Assuredly. One such change was the increased reliance on multiple-choice exams. But I wish to focus on another change, demographics in American universities.

This was the role of Asian-Americans in student bodies. I was able to witness, at the waning of the twentieth century, that their contingent, however numerically small, took the uppermost ranks in even elite institutions such as Ivy League colleges. Typically, these students were born in the US to immigrant families. Their foreign-born parents imbued them with the ethics of discipline and hard work—not to mention enrolling them in a host of extra-curricular activities, such as playing a musical instrument.

A few years, though, were enough for American values to start impinging upon the Asian mindset. As I mention elsewhere in this book, as a formal retiree, I was lucky to be called upon to serve visiting professorships at Cornell in 2000 and 2001. I taught there an undergraduate course, Chemistry 106, designed for non-science majors who had to satisfy a university-wide requirement—for graduation—to have taken at least one scientific course. The distribution of grades for that course was bimodal—the bulk was average or mediocre but a small contingent, 15–20% of the total, scored highly, showing that they had been interested and had studied earnestly. Asian-Americans were a significant part of that minority. But that is not the only gist of the story. I recall one Asian-American student asking me repeatedly if I was curve-grading: he had figured out, or so he thought, the system and he believed that he knew how to beat it, i.e., getting ahead with the minimum effort.

A similar attitude towards multiple-choice testing characterized student bodies in both Europe and America during my career: getting a good grade from being good at guessing which obviated the need to master the underlying body of knowledge. It takes quite a bit of expertise to devise a multiple-choice question that is not transparently easy to answer!

To me, the value of oral exams was the one-on-one, face-to-face interaction: I always strove to turn an exam into yet another teaching opportunity.

Roving and Foraging Widely: Or Wildly?   15

The words "curious" and "curiosity" come from Latin. They derive from *cura*, i.e. care; from which also came "cure" (as in a medical cure) or the French *incurie* (carelessness). These words indeed conveyed some denigration: the Latin word *curiosus* also termed a spy! Curiosity thus until relatively recently sustained a somewhat negative connotation. The Scientific Revolution brought a turnaround. There were harbingers of the change, though. Montaigne (1533–1592) declared (*Essais*, I 27, 1580) that "Glory and curiosity are the two plagues of our soul. The one leads us to stick our noses everywhere, and the other forbids us to leave anything unresolved and undecided." But not very much later (by 1605), Francis Bacon (1561–1626), in *The Advancement of Learning*, could write that "men have entered into a desire of learning and knowledge, sometimes upon a natural curiosity and inquisitive appetite."

I must have been graced with both curiosity and an inquisitive mindset. Hence, my turning into an experimental scientist. This chapter will evoke a dozen of my explorations of chemistry—taking the field as a whole.

As my Princeton appointment was getting to a close, in 1969, Edward M. (Ed) Engler (b 1947) and I scrutinized the interaction of camphor with benzene molecules in the liquid state, camphor being dissolved in benzene. Our query was how to best describe their interaction. It was the summer. My aim was the root cause of observations on molecules in solvents such as benzene—so called aromatics. The test molecule, ideal for a number of reasons, was camphor. I entrusted the experimental work at that time to Ed Engler, a young graduate student, who had joined my group at Princeton the previous year. He was a hard, dedicated worker. Ed and I got along fine, in spite of barely understanding one another, with my French accent and his to me opaque Brooklyn accent. Ed, following my instructions, achieved a large corpus of measurements. I pored over them, giving them close scrutiny. I did not want to prejudge the interpretation with pre-existing models from the literature, but instead to assemble it gradually, from scratch so to say.

What strikes me most in rereading our article is how French it is: in logic and clarity, not so much in the write-up but in the designing of the data collection. We

P. Laszlo, *A Life and Career in Chemistry*,
https://doi.org/10.1007/978-3-030-82393-1_15

started from a tabula rasa, with a deliberate absence of preconceptions. Ed's contribution was impressive, a wealth of reliable measurements. In addition to their analysis—orderly indeed—my input came from my familiarity with a large body of literature. One of the footnotes had no fewer than 66 entries, alphabetically arranged. My taste for history of science showed already then (1970) in citing a 1908 paper by Dolezalek. Why did I quote it? Because it presented what was in my eye an equally ridiculous viewpoint as the butt of our criticism. Our paper had also a polemical intent, showing that the model resorted to by a vast majority of authors was inadequate. It was a me against everyone else quixotic attitude and, I am sorry to say, the paper has a definite sermonizing tone: surely I was influenced in that by my Princeton colleague Kurt Mislow.

The study belonged to *molecular physics* more than to chemistry proper. If it had importance, it was because it taught us, Ed and me, how to frame, observe and report a phenomenon.

This was the summer of 1969. One underlying issue was payment of Ed's salary during those months. I arranged a donation with Mr. Murphy—at the time executive assistant to the chairman of the Department, Kurt Mislow (1923–2017). My salary thus served to support Ed Engler during those months.

This research turned out beautifully. The ensuing article was completed by early spring in 1970. It was accepted in *JACS*, opening the issue of March 24, 1971. Our results ruled out a classical but inadequate model of one-to-one complex formation between camphor and an individual solvent molecule. Instead, a time-averaged cluster of solvent molecules around dissolved camphor gave a much superior and realistic interpretation of the extensive data—we made measurements in no fewer than 50 different solvents; our article of over 11 printed pages and nearly as many data plots was rather lengthy, rich in detail. I remain to this day proud of our contribution in that article. It has robustly withstood the test of time.

If this was an incursion in *molecular physics*, the next notesworthy achievement from my lab—now in Liège, I started there at the beginning of 1970—involved André Cornélis (1941–2021), one of my first two Belgian graduate students. It belonged with *mechanistic organic chemistry*—i.e. the step-by-step monitoring of a transformation. For his Ph. D. I assigned André to work on the close and careful following of the interaction of two molecules, one with a legendary avidity for electrons (known as tetracyanoethylene), the other electron-rich (known as dicyclopropylfulvene). What was their embrace like? What would it lead to?

We were mystified—what can I say, we were bowled over, utterly flabbergasted—by the actual outcome. It turned out to be a fascinating set of events, at the atomic scale—nano as by now it is customary to say. One anticipated the two reactant molecules to embrace tightly so that electrons could flow from one to the other. And indeed a so-called Diels-Alder transformation took place with formation of a bicyclic adduct. However, this initial product was unstable and it rearranged into another also bicyclic product, this time with a cyclobutane C_4 ring. Yet, this was not the end of the story of the sequence of transformations. The cyclobutane intermediate suffered considerable strain and released it by isomerizing in turn to a final product with two C_5 rings.

We documented in detail these transformations from plots of the reaction rate versus characteristics of the various solvents used: their dielectric constant gave us a handle onto electrostatics in the transformations, their polarity and nucleophilicity informed us as to what occurred, i.e., the rationale behind these two successive isomerizations. Ever since I took in 1958 or so a class taught by Noël Félici (1916–2010), a brilliant scientist and teacher, electrostatics have fascinated me. As we wrote in the Letter to the Editor of *JACS* in which we reported these results, "the tetracyanoethylene residue moved from its initial position above the face of the fulvene, allowing for maximum overlap and thus producing the fastest rate, to its final resting position in the nodal plane of the fulvene, where strain has been reduced to a minimum."

From elucidating the intriguing details of a reaction in organic chemistry I moved into a novel area, of intense activity at the time, *organometallic chemistry*. The reasons were predominantly a chemist of genius plus two talented coworkers.

The genius was Rowland Pettit (1927–1981), a professor at the University of Texas in Austin. He came to Princeton for a lecture series—Kurt Mislow, chairman of the department at the time, tried unsuccessfully to lure him into the department—he also tried and failed with Jack Halpern (1925–2018), from the University of Chicago, another bright star in organometallic chemistry. Pettit was a shooting star, his cigarette addiction killed him all too early—he was only 54—, as it did many others from lung cancer.

I was entranced by the beauty of the transformations Pettit showed us. Often they involved metal carbonyls. I resolved to find out what they could achieve. I convinced the most senior graduate student in my group, Arnie Speert (b 1945), to give it a try. He did and I was hooked. But Arnie could not carry this project further, in 1970 he got his Ph.D. and left Princeton. This was also the year I began in Liège. This new line of research had become dear to me from the elegant directness of the transformations. I assigned a follow-up to Armel Stockis, a Belgian graduate student—and his results were stupendous: reaction of an ethylenic molecule with an iron carbonyl could have formed 36 different isomers—but it gave us only one. Such selectivity rivalled that of biotransforms. Armel and I then fielded the tough but intellectually challenging assignment of identifying which of the 36 possibilities had occurred. This structural identification required the whole repertory of the then available physical methods. Armel, by succeeding at it, achieved a masterpiece comparable to that of a craftsman of yesteryear, having assembled an intricate and creatively novel design.

From the vantage point of the decades elapsed since publication in 1974 of our Letter to the Editor of *JACS*, I see that our all-too-brief communication ought to have been followed by a full paper, to do full justice to the beauty of the work Armel had carried out nearly single-handedly. It reads like a mathematical demonstration, razor-sharp. True, at the time, Armel was a man of few words, he projected a Buster Keaton-like personality. What our paper lacked is the self-promotion, for instance at least a reference to the *Organic Spectroscopy* book I had co-authored with Peter Stang. Establishing which of the 36 different stereoisomers was the reaction product was a small feat of logic and elegance, which only expert stereochemists were able to grasp at the time.

My appointment at the University of Liège brought to me pros and cons. The negatives were the second-rate status of the institution, especially compared to Princeton; the mediocrity of many of my colleagues and the ensuing lack of competition and stimulation. There were positives though: my youth, the status as a full-professor and the anticipation by some onlookers, at the Free University in Brussels in particular, that I would nevertheless succeed. I told myself and the coworkers in my research group that we had an unassailable position, by being at the very bottom of a well. Hence, the only direction we could go was up.

At the end of the summer of 1973, I went to Brittany to attend a conference of top-of-the-line organic chemists there. This was GECO, the fourteenth in a series of yearly meetings began in 1959 by Guy Ourisson, on the model of the American Gordon Research conferences. Whom do I recall among the attendees, about 40 people in total? Andrée Marquet (b 1934), a French lady-chemist of note; Dieter Seebach (b 1937), a German rising star who would receive, 4 years hence, the call to a chair at the world-famous Swiss Polytechnic in Zürich; there were also two leading American chemists, Ron Breslow (1931–2017) of Columbia University and the already mentioned Kurt Mislow who had been my chairman at Princeton. Breslow and Mislow both belonged in their country to the prestigious National Academy of Sciences.

I presented to the group this work of Armel's. The presentation brought me applause and high compliments. Mislow, an expert in stereochemistry, commented that the selectivity I had reported was a record high among non-biological transformations.

And myself? This was a turning point in my career. It flashed through my head that my desert crossing that began with my appointment in Liège had ended. I had joined the club of scientists enjoying high-regard from their peers. But it did not end there. As the forthcoming pages and chapters will tell, a chapter in my career ended and another one would open shortly. Prior to 1973 and the conference in Locguénolé, my contributions amounted to vignettes—each such miniature exquisitely drawn and a kind of perfection in itself, but still a miniature and not entry into a whole new area.

I was prepared to jump at such an opportunity if and when it came along. As I shall report henceforth, it came my way not once but twice. However and prior to recounting these two chapters of my scientific life, I shall briefly mention some of the directions in which my meanwhile relentless curiosity took me within the chemistry fold.

In 1970 Raymond Dratler and I became interested in a solid semiconductor, made of iodine and a nitrogen-containing partner molecule, pyridazine: *solid-state physical chemistry*.

We, i.e., Jean Grandjean (b 1945), Michel Jauquet and I investigated in 1972–1973 the molecular dynamics of camphor in solution: this globular molecule rotates upon itself: a study in *molecular physics*.

Also in 1973 Ève Marchal, Agnès Paris and I reported the effect of acids on a nematic mesophase, a study in *physical chemistry of polymers*.

During the late 1970s, Jean Grandjean and I performed *biophysical studies* on a ubiquitous and metabolically all-important biomolecule, calmodulin, a calcium-binding protein.

Alfred Delville also took part in that investigation. Another related piece of research involved him, during the early 1980s. It was the question (*physical chemistry*) of whether ions bind at specific locations or just congregate around polyelectrolytes in solution.

And there were quite a few other studies on rather diverse topics, that all more or less qualified as chemistry, between the two poles of biology and physics. In 1973, a joint study with colleagues in the School of Medicine of the toxicity of a couple of chemicals towards specific cells, in conjunction with grafts rejection by the body. Experimental determination of the electronic distribution in sulfur-containing molecules in 1976. Interaction with neighboring molecules of the thyroid hormones known as T3 and T4, also in 1976 (*biophysics*). Electronic transitions of the tri-iodide anion, still in 1976 (*physical chemistry*).

The first opportunity for a major contribution came my way as early as 1976. I seized it. For several years, I lived the thrilling adventure of scientific discovery. It was mind-boggling, it was engaging my attention at every instant and, needless to say, it was highly competitive. This was the phenomenon of the self-assembly of 5′-GMP—to give it its technical name.

How did it begin? From a small, nearly unnoticeable discrepancy. As part of our biophysical investigations, I suggested to Agnès Paris (b 1950) she study the workings of the enzyme, ribonuclease, with a nucleotide molecule—this 5′-GMP acronym—as a substrate mimic. She started doing so and plotted the results. The graph did not quite extrapolate through the origin, as it should have. I did not ignore this tiny anomaly. I guessed that the explanation was self-association of this nucleotide.

Not only that. Intuition is a very opaque manifestation of the mind. I knew somehow and from the offset that this was an important problem. I became obsessed with it for months on end, even years. A sound sleeper as a rule, I went through night-long insomnias. To bolster that intuition, competition set in. A research group at the NIH, in Bethesda near Washington DC, headed by H. Todd Miles, was hard at work in the attempt to decipher this self-assembly. They published a communication in the November 1 1975 issue of *JACS*. This was formidable competition. It did not stifle us, to the contrary it made us even more desirous to excel; a most fruitful rivalry—it taught me how much scientific advances thrive on controversy and indeed competition to be the first in reporting secure knowledge.

After confirmation that self-association was indeed taking place, I had Agnès monitor this phenomenon with a tool our group had recently honed, nmr of the Na^+ sodium ion. This was during the fall and winter of 1975 and the early spring of 1976. I reported these results and what they showed—an aggregation number of 16–20—at the fall 1976 meeting of the American Chemical Society in San Francisco. As I noted in the write-up (pre-publication is dated to June 1 1976), "these phenomena may have considerable biological significance."

By that time, I knew in my bones that the problem was indeed of prime importance. While Agnès continued to work on it, I directed others in our research group to its further study. These were Marie Borzo, a post-doc from Colorado State University who had come to spend a year in Liège; Alfred (Freddy) Delville, a young mathematician whom I had recruited to lead our efforts at modelization; and Christian Detellier (b 1952) a fantastically-gifted graduate student, who indeed bore the brunt of this difficult but highly significant research.

Yes, it occupied me single-mindedly for many days and sleepless nights. I started also, surely also because of the competition with the group at NIH, to report our advances in lectures and seminars. One such took place at the University of Connecticut, in Storrs, where I had been invited as a visiting professor in the fall of 1977. There I shared details of the story, as it had unfolded so far, with my host, Professor Edward T. Samulski. He was a good friend ever since our joint days at Princeton, where I was a faculty member and he a graduate student in polymer chemistry, whom I did a good turn to when he failed to satisfy the cumulative exams requirement. My colleague John M. Deutch, later director of the CIA, forcefully wanted him expelled. But I knew Ed to be exceptionally gifted and creative.

He displayed that very creativity in his mailing to me, from Storrs to Liège, a homebuilt molecular model (Ed is a great tinkerer) accompanied by an explanatory letter. Ed proposed thus a structure, or rather a set of structures, for the self-assemblies we were so intent upon understanding. It was a splendidly conjectured architecture, the harbinger of very many reports—still in the distant future.

My enthusiastic dive into the wonderland of self-assembly by a nucleotide, before I tell of our key results and their biological significance, brings up the topic of intuition. Ought a scientist resist it or conversely follow it? If the latter, why is it important? Is intuition contrary or just complementary to rational analysis? These are a few relevant questions.

I followed my intuition out of a powerful gut feeling it was the right thing to do. I felt encouraged in doing so, firstly by what my father once told me, when I was turning from adolescence to young adulthood, "your imagination will save you." Imagination is not intuition, but they are related. Another statement that made also a deep impression, a few years later while listening to a presentation by George Whitesides (b 1939) in a conference was hearing him say, initiating an answer to a question, "my intuition tells me . . .".

But what indeed is intuition? Etymologically, the word derives from the Latin verb *intueri*, with the meaning of "to watch attentively, to consider fixedly". More or less during the seventeenth century, this meaning that had been embraced by the scholastics was replaced by the meaning of "an immediate apprehension." Which in turn, within modern philosophy, switched to "an apprehension either mediated by the mind or by the senses." Wittgenstein, however and I follow him on this regard (as in others) thought that our possession of intuitive knowledge is a necessary truth.

In addition to Agnès, a most reliable experimentalist and a keen observer, my coworkers included also—for modelization purposes—Freddy Delville, a promising applied mathematician, and Christian Detellier (b 1949), dedicated and enthusiastic, who shared my thirst for understanding the self-assembled structures this nucleotide formed in solution. My role was akin to that of the leader in a chamber music ensemble, i.e., the weaving together of their individual contributions; and also, needless to say, as a writer. The former translated into devising tactics of elucidation. The latter meant writing up reports that grew together with each of our advances in understanding. We could not wait for it to be complete, competition with the group at NIH dictated such gradual expression.

During the period 1976–1982, we published a dozen papers on the 5'-GMP-self-assembly. A key report of ours was nucleation by sodium and/or potassium ions of the stacking of G_4 tetramers into higher-order aggregates such as octamers and dodecamers. The big prize, though, regarding the major biological significance of such aggregates came in 1988 from Walter Gilbert (b 1932) at Harvard. He was the 1980 Nobel prizewinner in chemistry. He discerned and established the relevance of G self-assembly to meïosis, i.e. cell division: the terminal ends of chromosomes—the so-called telomeric ends—are guanine-rich, which provides the requisite glue for their binding at that stage. Again with D. Sen, Gilbert reported in 1990 a sodium-potassium switch in the formation of four-stranded G_4-DNA—which our 1982 paper with Christian Detellier had anticipated. In their paper, Sen and Gilbert referred to the work by the NIH group of Miles, Becker and Pinnavaia but most regrettably not to the rival achievements by our group in Liège. I shall come back to the aftermath of the G_4 self-assembly in a subsequent chapter. Enough to say here that my intuition in 1976 of its key importance was to be abundantly vindicated.

Even though I was only 18 and a young student in chemistry, I still vividly recall the potter Norbert Pierlot's (1919–1979) insistence "you ought to study the chemistry of clays." This was a visionary statement, it took me only a quarter of a century to heed it. To preface this other intuition of mine, that begat a big and highly productive research effort, I need to refer to a great scientist although also a doctrinaire Marxist, the Englisman J. Desmond Bernal (1901–1971)—both Pierlot and Bernal died too young, they were only 60. In 1967, Bernal published a book in which he made a strong case for the role of clays in the origin of life on Earth. To quote André Brack's (b 1938) useful summary of Bernal's argument (2006), "the advantageous features of clays are (i) their ordered arrangement, (ii) their large adsorption capacity, (iii) their shielding against sunlight (ultraviolet radiation), (iv) their ability to concentrate organic chemicals, and (v) their ability to serve as polymerization templates." Bernal's views influenced me deeply and I was all prepared to start experimenting along them.

Now to clays in chemistry. How did I intuit the potential of these minerals to chemists? In the mid-1970s, Keith Pannell (b 1940) came to my laboratory in Liège. He had learned we were active in sodium-23 nmr. The idea of using clays to safely store radioactive waste is straightforward: if one sets a heavy metal, an uranium salt for instance, U^{6+}, in presence of a clay, it migrates inside the clay displacing in the process half-a-dozen univalent ion such as sodium and potassium. Once the heavy metal sets there, it will stay put. Keith's notion, it had led him to come all the way to Belgium from the University of Texas in El Paso, was thus to simulate the behavior of uranium salts with other heavy metals and monitor the ensuing leaching out of univalent ions using sodium-23 nmr. It worked fine and Dr. Pannell returned to the US with a trove of gratifying results. In the meanwhile, he had become a friend. Already a well-established scientist at the time, he remained very young in spirit. A seasoned traveler, he roamed the planet with the lightest luggage and went wherever his fancy drew him. He was very much an adventurer. I liked him. It reinforced my notion of what the atmosphere in a research group ought to be: an aspect "camp of nomads, of hunters-gatherers."

The story then took a different turn. A student in Liège had to prepare her senior thesis. Christian Detellier, fresh from his own Ph. D. dissertation, asked my permission to supervise her work—which I agreed to willingly. Christian is enthusiastic, commits himself totally to everything he does, is generous, loquacious and funny, very smart and attentive; assets that worked wonders in our joint work. I still feel privileged to have been present at the birth and nurture of this great scientist. Christian set the work of the young intern as a topic in prebiotic chemistry—to find a role for clays at the origin of life, no less. He chose to study transformation of a nucleoside into a nucleotide. The switch in a single consonant indicates a more profound chemical change. Anyway, this research work met with full success. However, we never published it. I was uneasy about inferring any conclusion about the origin of life, because millions of scenarios can be entertained. I felt on a slippery slope—that from science to science-fiction. This trouble lasted only a month or two. I woke up one morning with an illumination: if clays could have assisted a reaction of prebiotic significance, they might be useful in everyday chemistry, of the kind that is the daily fare of laboratories.

That the brain during sleep comes up with intuitions, sometimes significant ones, has been abundantly documented. This particular one came as an analogy, a familiar trope in my thinking.

André Cornélis, who had assisted me since 1970, then took over. His very first attempt—modelled after a procedure known as phase-transfer catalysis—was brilliantly successful. A whole new chapter opened itself to us, transformations in organic chemistry using clays to make them more effective: faster, high-yield and selective. Keith Pannell had left us the collection of clays he had brought with him. Even though we never used them, it gave us confidence: acquisition of his sample case somehow prompted us to acquire the knowledge we lacked.

Freddy Delville initiated sophisticated calculations about electrically charged particles (ions) interacting with clay surfaces. And Balogh Maria (b 1949), a highly experienced Hungarian synthetic organic chemist, quietly but most efficiently started running reactions in the presence of clays.

It was magical. The magic of discovery. The magic of apparent catalysis—in fact due to the increased probability of encounters between reaction partners on the surface of the flat clays platelets, as compared to the surrounding bulk liquid, a surface compared to a volume. It was a magnificent scientific adventure, endlessly stimulating to further advances and to learn yet more.

It was rewarding too. We started publishing this work in 1980. By the end of the century when I took early retirement, no fewer than 50 primary papers reported our work on chemistry using clays. Some of the reagents we had devised entered the chemical literature with their acronyms, *clayfen* (clay-supported ferric nitrate), *claycop* (clay-supported cupric nitrate), *clayzic* (clay-supported zinc chloride), etc. In a following chapter, I shall examine the medium-term aftermath of the above-related intuitions. Of course, I wont claim having been a visionary, only a mover: I believe a scientist has a moral obligation of moving along the sum total of knowledge.

Visiting Professorships

To me, a visiting professorship is an integral part of the life of an academic. The transplantation immunizes against sclerosis, from staying too long in a particular atmosphere. To breathe another is essential to one's liveliness and creativity. This was not only instinctive on my part, it was also a lesson from the life of a humanist, such as Erasmus (Fig. 16.1).

In addition, I needed the intellectual stimulation that was almost totally absent from my permanent appointments, in the chemistry departments at the University of Liège (1970–1999) and at the *École polytechnique* (1986–1999). Hence, seeking it elsewhere, on the run so to say.

A major factor in my acceptance of visiting professorships was the challenge and the ensuing satisfaction of becoming fully operational in a foreign and usually somewhat exotic locale. I had a rule though: never to ask or apply for such an appointment but rather wait to receive an invitation to serve in such a manner.

As I wrote above, to me it is an integral part of being a professor. With Hungarian forebears, I must have a bit of the Gypsy in my blood! When I read of the Hungarian mathematician Paul Erdös (1913–1996), whose belongings would fit in a suitcase, as dictated by his itinerant lifestyle, I admired him as somewhat of a model.

In addition, I have remained somewhat of a little boy and have always been pleasurably excited from stepping into a train or an airplane. I relish having to myself the time of the trip. I have written many a paper in such circumstances. The best way I can convey the feeling travel gives me is as a conduit for communion with others. When I read Bruce Chatwin's *Songlines* (1987), the nourishment he received from the Australian landscape as from an umbilical cord was familiar to me: "each totemic ancestor, while traveling through the country, was thought to have scattered a trail of words and musical notes along the line of his footprints."

Was it Tuesday, April 30 1968? I was in the *Capitole*, the fast train that ran from Paris to Toulouse. There was just one other person in the compartment. This gentleman in a suit was reading a paper I had never heard of, entitled *Minute*. There was a big headline, *Qu'on expulse l'apatride Cohn-Bendit* (Expel the stateless Cohn-Bendit). I had no idea what it was about but I was to find out all-too-soon.

P. Laszlo, *A Life and Career in Chemistry*,
https://doi.org/10.1007/978-3-030-82393-1_16

Fig. 16.1 My Cornell ID, after some years! (from the author's wallet)

I was on my way to a visiting professorship in Toulouse, for the entire month of May. I had flown over from the US to take it on. This was at the invitation of the Faculty of Sciences of the University. It had its lecture rooms and laboratories in the suburb of Rangueil. I was a guest at the Institute of Biochemistry. My hosts there were two colleagues, Jean Asselineau (1921–2013) and Jean-Pierre Zalta (1924–2015). A few days later my wife Martine and daughter Sophie, then 4, joined me. We stayed for the whole month in the Asselineau house, at the invitation of Jean and his wife Cécile Asselineau (1921–2020)—two lovely persons, with a son who was about 10.

Wednesday May first was Labor Day. I may have given the first lecture I had prepared on Friday May third, at least that is my recollection. But I never gave lecture Number 2. Over the weekend, the entire university went on strike. This was the beginning of the May 1968 events in Toulouse.

They had started at the new university of Nanterre, near Paris. The Minister of Education, Alain Peyrefitte (1925–1999), was visiting. Representatives of the students, led by Daniel Cohn-Bendit (b 1945), presented him with some requests, such as easier access to the women's dormitory on evenings. To which the minister replied, in jest, "jump in the pool to cool yourself off, young man." Which started a strike by the students locally; which then spread to most universities in the country. About mid-May, the entire country was on strike, paralysed.

Meanwhile, the university system underwent occupation by the students. I witnessed it in Toulouse. Lecture halls, in each of the faculties, were filled to capacity. Discussions went on, all day long. The administration had fled. The Recteur—i.e., the President of the University—and the Dean of Sciences showed the utmost caution and were both secluded in their homes.

What was being discussed, endlessly? Desirable reforms. Why endlessly? Because of the French, cartesian mindset: one has first, to agree on principles. Which obviously was not an easy goal to meet.

Day after day I joined my colleagues, Asselineau and Zalta, in the lecture hall where students and faculty from the Sciences congregated. A small group of faculty members saved the day, by their presence and at least passive participation. They avoided destruction of the teaching and research tools. These were interesting times. Among other things, I witnessed a power takeover attempt by a young colleague in chemistry, in Napoleonic style.

At the end of the month, it became impossible to continue in this manner. With the postal system on strike, I was totally cut-off from my research group at Princeton. On Friday May 30th, the three of us rented a car, made the rounds of service stations to fill it up with gasoline—it was becoming scarce—and drove to Switzerland. It was eerie, the roads were deserted. We saw only a few *gendarmes* patrolling the countryside. France was scared. That evening, De Gaulle made a speech and it caused a thaw, normalcy gradually returned.

Upon my own return to the chemistry department at Princeton, I went to see my chairman, Kurt Mislow. I spent an hour narrating what I had seen and heard. He listened carefully, took some notes. He then went into action, setting-up joint committees with students in the department: American efficiency at its best. In a couple of days, more was achieved than in France during the whole month.

That was my experience of the Students Revolution of 1968.

It may have been during the Bürgenstock conference on stereochemistry in 1974 that Manfred Schlosser (1934–2013) invited me to a visiting professorship at the University of Lausanne, where he taught. I believe 1976 to have been when I did so.

I nurture endearing memories of my attending stays in Lausanne, a city I cherish. At that time, the University was still located in the old part, next to the cathedral. The Institute of Organic Chemistry, where I had a desk, was at number 2, rue de la Barre: from my room at Hotel Crystal on rue Chaucrau, after crossing place de la Riponne, I had to ascend ensuing steep staircases to reach it, a 10-min walk. I would be greeted there by another Frenchman of my age, Jean-Pierre Kintzinger (b 1942). A former graduate student with Jean-Marie Lehn in Strasbourg, he was an endearing Alsatian: warm and congenial, always smiling, bright and sunny in his attitude to life and an enthusiastic research scientist.

Manfred Schlosser was a more taciturn character, very self-enclosed—self-congratulatory to an amazing extent. He was into organometallics, had studied with Georg Wittig (1897–1987) in Heidelberg and had remained a specialist of the Wittig name reaction. Manfred and his wife, also German in my recollection, had three children, Anya, Katia and Ralph—thus named, it was said, for the three types of intermediates in organic reactions, anions, cations and radicals.

Other colleagues I interacted with while in Lausanne were Hans Dahn (1919–2019) who administratively headed the Institute of Organic Chemistry, he owned an impressive personal collection of kilims; Pierre Vogel (b 1944), from a Vaudois family of vintners; André Merbach (b 1940) in analytical and inorganic chemistry; and, in later years, Geoffrey Bodenhausen (b 1951) in the methodology of nmr.

Life in Lausanne? What I enjoyed most was riding the *Métro* up from the lakeshore to the center of the city; visiting the Payot bookstore, it stood then at a bottom corner of Rue du Bourg; stopping on that street for a cup of coffee and a pastry; repairing for dinner in one of the old-style *carnotzets* (wine-tasting cafés); shopping at a Migros store—at Jean-Pierre's suggestion I had enlisted in that cooperative; opening a numbered Swiss account in a local bank—I held on to it for many years afterwards.

I enjoyed also and very much skiing excursions with the entire Institute, once into the Bernese Oberland, another time in the huge fields across the border with France, named *Portes du Soleil* (overhanging, in France, Morzine and Avoriaz).

A recollection I shall always miss is Jean-Pierre's joy at being a scientist. He was clearly on a winning trajectory. However, he was compelled by the serious health condition of a daughter to return to Strasbourg. In doing so, he had to retire from active research and became a university administrator; it was a pity, he was a superb scientist on the rise.

I have saved this anecdote on life in French-speaking Swizerland for the last. Martine my wife had come to visit for a weekend. We went skiing together in Zermatt. On the Monday, we got on a train back to Lausanne, just in time for delivery of my lecture in the Institute at the end of the afternoon. I had forgotten my lecture notes inside the little shuttle going from the Alpine resort down to Visp in the Valais, where one transfers to the main line. Martine became panicked and suggested we flee and go home. I tried to pacify her. I was able to reconstitute most of the lecture from memory. At the end of the evening, after having established by phone safe retrieval of my handwritten notes, we went down to the railroad station to recover them. It being Switzerland, I paid a small reward—20 francs, if I recall—to the employee who had found them.

I met C.H. (Chuck) DePuy (1927–2013) at a conference. He invited me to visit at the University of Colorado, in Boulder. This came about during the summer of 1983. I was thrilled to spend time in this endearing mountain community. I had a room in a dormitory. Each morning, I would walk about a mile one way to purchase *croissants* in a bakery I had discovered—they were the best I have ever enjoyed anywhere. I saw there for the first time students going to classes wearing backpacks—a fashion that would engulf the planet. Boulder sits at an elevation of a mile, which suited me fine. Each day, I would go further up a hill in midtown—a hike of between an hour and two round-trip.

For instance, on Saturday August 6 we drove with Eleanor and Chuck DePuy and left the car at the bottom of Gregory Canyon. A trail follows that little valley and goes up towards Flagstaff Hill. After walking up together for about 500 m they suggested I continue by myself—towards Green Mountain, behind the Flatiron rocks. Which I did, reaching the top (8100 ft) around noon. I met there Marion Trifunac's Dad, a retired professor of electrical engineering at the University of Colorado. I then walked down, leisurely, from the mountain. While doing so, I made a somewhat eerie encounter: a glider flew just above where I stood, fewer than 50 m away. The noise it made was not unlike that of electricity in a wire.

Chemistry-wise, the University of Colorado was a treat, on account of its organic chemists. There was Stan Cristol (1916–2008). He was a member of the National

Academy of Sciences and a brilliant physical organic chemist, most accomplished in elucidation of reaction mechanisms.

When he and I talked, he told me of photochemical work in the bicyclic series; of how, in particular, electron transfer occurs to an *anti* C-Cl bond; together with migration of a ring: the ring that does not donate the electron is the migrating ring.

Stan Cristol was a great leader, he brought eminence to the University of Colorado. Chuck DePuy had a different personality, he was a brilliant, an outstanding loner. Physically towering, at six foot five, he had a superb intellect. His focus then was on gas-phase reactions, work that earned him fame (too much perhaps) in his lifetime. Our interaction was of few words, when I visited him in his home—his wife was then absent. Chuck was immersed in reading the *Science* weekly. Once, he gave me this compliment: like him with our work on clays I favored the off-beaten track.

From my time in Boulder, I recall a piece of gossip by the junior faculty in chemistry. They commented the invitation Harvard University had just made to Stuart L. Schreiber (b 1956), especially regarding the lavish equipment of his laboratory (he would start there in 1988, some years later). I remember being appalled by the tone of the whole exchange, envy at its worst.

That summer of 1983, in addition to Dave Walba—more on him later—I interacted predominantly with two assistant professors, Gary Molander (b 1953) and Christopher S. (Chris) Shiner. Their then parallel but ultimately divergent paths made a study in contrasts.

The latter had everything going for him—seemingly. He was part of the inner circle, with a family connection to a professor at Columbia: he had prepared his Ph. D. with E. J. Corey at Harvard and been a post-doc with Gilbert Stork at Columbia. Stork was his father-in-law, he had married Linda Stork, who was an MD (pediatrics). This patrician was a self-appointed œnologist and expert in wines.

Chris had a critical mind—way too acidic for his own good and it may have caused his downfall. He had obtained three research grants, after writing eight proposals—not a bad record. At that time, the University of Colorado deducted 40% overhead on outside grants. Hosting a postdoc would thus cost $25,000 to a group leader. He criticized a fellow assistant professor, David Walba, for not being strict enough with his coworkers. He criticized also Kishi at Harvard, one of the three successors to R. B. Woodward, for being too much of an empiricist and making graduate students in his group try anything however outlandish.

Chris worked on hydroboration, attempting to remedy H. C. Brown's methodology where it was deficient—the critical mind again. He worked also on a synthesis of thromboxane A2.

The more plebeian Gary Molander also devised novel synthetic methodologies. As soon as I had met the two of them, I was confident I knew whom would succeed. Molander was philosophical about failure or success in the lab. As he put it: "what you strive for is not what you get." And indeed my hunch was verified by the outcome in both careers. Shiner did not obtain tenure at the University of Colorado. He dropped out of academia. He and his wife Linda settled in Portland, Oregon, where he opened a chemical consultancy. Molander, conversely, enjoyed a fine career: tenured at Colorado in 1988, and called to the University of Pennsylvania in 1999; very productive, in the devising of new synthetic methodologies.

David M. (Dave) Walba (b 1949) was impressive, I felt. Seemingly eccentric, he followed his own taste rather than bowing to the trends then current. He was interested only in stereochemistry, in its most recondite corners—such as molecular knots. He used graph theory, then the province of very few chemists. He has had a very productive career, along these very lines, entirely at the University of Colorado.

While in Boulder, Robert G. (Bob) Bergman (b 1942) came from UC Berkeley to visit and give a seminar—on August 5 1983. He was very impressive. He talked about activation of carbon-hydrogen bonds using transition metals such as iridium. That Sunday, our hosts took us, Bob and me, on a drive through the Rockies, to admire Pikes Peak in particular.

I first met Armin de Meijere (b 1939) at the University of Göttingen in the late 1970s, he was a *privatdozent* there. We met again at the Bürgenstock conference on stereochemistry and I had invited him to present a seminar in Liège; which he did on June ninth, 1983. On that occasion, he invited me to serve a visiting professorship at the University of Hamburg, where he then taught and had his laboratory.

This I did during the spring and summer of the following year, 1984. I split my time between Liège and Hamburg, thus discharging my responsibilities in Liège as a professor and laboratory head and giving a lecture series in Hamburg. The commuting was easy: I would get on a train in Liège and transfer in Köln to a northbound fast train, going through Hannover and two or three other stops. The whole trip took half-a-dozen hours under comfortable conditions, I was able to read and write. In Hamburg, Armin had secured use of an apartment for me in a suburb, reachable by suburban train. This furnished flat belonged to a former coworker of his who had left for a position in chemical industry—with Hoechst in Frankfurt is my recollection.

The Institute of Organic Chemistry of the University of Hamburg, at Martin-Luther-Platz, was within walking distance of the main railroad station. It was headed by Professor Wolfgang Walter (1919–2007), a courteous gentleman nearing retirement, 2 years hence.

Armin was a livewire. He spent relatively little time in Hamburg, intent upon professional advancement. He courted Princeton, where he would spend part of 1985 on a visiting professorship. Ultimately, he received a coveted call from the University of Göttingen where he would spend the rest of his career, from 1989 until 2006.

One of his coworkers, whom I met in Hamburg, was Joe Richmond (b 1934), a remarkable American-born and educated chemist who went into scientific publishing, working for Thieme Verlag, Stuttgart from 1986 to 1995 and for Springer Verlag in Heidelberg from 1996 to 1999—afterwards retiring as an independent publishing editor.

Another associate of Armin's I got to know and befriend, together with his wife Angelika, was Dieter Kaufmann (b 1948): in 1984, he was working toward his Habilitation. He also became a full professor, called that he was in 1993 at the Technical University in Clausthal, in the Harz mountains.

Among the colleagues of Professors de Meijere and Walter in Hamburg whom I came to know and respect, another two deserve mention, Dieter Rehder (b 1941) and Ernst Schaumann (b 1943). The former, who spent his entire career at the University of Hamburg, became the world expert in the bio-inorganic chemistry of vanadium, studied by nmr. The latter went to the Technical University in Clausthal in 1990 and was responsible for the hiring of Dieter Kaufmann there, just a few years later.

In mid-June 1984, I attended Professor Schaumann's group seminar. The impressive opening presentation was by Hilde Nimmesgern, a coworker in the final year of her doctorate. She would complement it with a postdoctoral stay in Atlanta at Emory University, with Albert Padwa (b 1937). She then did research in medicinal chemistry at the Hoechst R&D in Frankfurt (1985–2000). And then when it became the German branch of the Sanofi-Aventis pharmaceutical company for 12 years (2000–2012), prior to becoming a senior business consultant in the Francfort area. Her's was a typical education and career for a German organic chemist of the time, which explains the detail given.

Professor Schaumann was vividly interesting. He belonged to one of the oldest families in Hamburg—but, at least this is my recollection—became estranged from his parents over the crimes enacted during the Nazi period. At that time of the late Eighties, there was a reckoning. In many a German family, the children confronted their parents over their behavior, their silence and inaction at that time. I was very much impressed with Ernst's integrity. He told this story about one of his forebears: this was in Napoleon's time. In 1813–1814, the French army in Hamburg was under siege. The French garrison of Hamburg, commanded by Marshal Davout (1770–1823), victoriously resisted for nearly 6 months to the coalized troops of Prussia, Russia and Sweden. At one point, Davout decided to expel from the city useless civilians, who died to a man. French soldiers demanded from Ernst's ancestor that he guide them to a ford on the Elbe river. At first, he balked; but did it after they threatened to kill a baby. This was during winter. Upon return, however, he got lost in a forest and was killed by the intense cold.

I loved the city of Hamburg. What did I like best? The inner lake, known as the Alster. Socially exclusive rowing clubs—Armin belonged to one—use it. It serves also for walks, hikes and jogs around it—which I indulged in. Hamburg used to be a major city and harbor of the Hanseatic League. It continued to be a very active port, near the mouth of the Elbe on the Baltic. At that time of the mid-Eighties, I would see *ro-ro* ships (roll on, roll out) moving in and out. These maritime activities brought with them a definite British influence, on social mores in particular. The attendant wealth flaunted itself in luxury shops in the center of town.

As a Frenchman, I was very much impressed by two features of German life: the intense Francophilia, radio programs very much dominated by French classical music; and the parenthood between the two cultures—it may help to explain why the two countries came to be so often at war—Germans rivalling us in logical and abstract thinking (think Kant and Goethe).

But my main attachment to life in Hamburg and the area came from Bach. On Sundays, I would repair to some of the tiny Protestant churches above the Elbe, where one of his cantatas would be played with a few instrumentalists and soloists: I had the impression of being transported back into the eighteenth century. Such a small ensemble re-enacting these pieces was extremely moving. I was also lucky enough to attend a presentation of *Matthäus-Passion* at Easter time in one of Hamburg's major churches. The music was great, but what sticks in my mind is the religious fervor of the audience: it too was a pilgrimage back in time.

During the summer of 1986, I answered an invitation from the University of Chicago. I had a room, with a Murphy bed, on top of an old building. The view was nice though, one could see trains arriving and departing from a major railroad station. My room saw some traffic as well, it was burglarized.

Getting acquainted, even a little, with the University of Chicago got me fascinated, it is an institution keen upon remaining different from others. As is my wont, I explored with awe the bookshelves in its library, the Ravenstein.

My job was not easy. It was lecturing to students who, as a rule, had already failed their qualifying exam in organic chemistry. They were jaded, they had no enthusiasm, they learned by rote—to me it was the worst possible audience. I did my best, though, to try and get them interested. The textbook was, at the time, the standard manual of organic chemistry by Morrison and Boyd (Chaps. 17–25, nearly 500 pages), that I found to be a dull, repetitive and turgid text. Hence, I had plenty of incentives for hard work, I hope I discharged it well. My lectures were scheduled three mornings a week, Mondays, Wednesdays and Fridays, 9:30–11:30—the break about 10:15 came none too early for both the students and myself.

There were compensations though. Tourism was one. I took advantage of the proximity to explore nearby parts of Wisconsin. I recollect my admiration at the fortitude of the pioneers who, many having come from Scandinavia, had endured harsh winters in their log cabins and managed to make a new life for themselves under trying circumstances, to say the least.

The geology of the terrain could be breathtaking, I remember in particular the lunar landscape of the Kettle Moraine area, a land formation pocked by retreating glaciers. The mosquitoes there were fierce.

In addition, there was in the chemistry department Philip E. (Phil) Eaton (b 1936) a towering figure of a colleague whom I admired and befriended. He had achieved this milestone in organic chemistry, synthesizing for the first time the elusive and seemingly unattainable cubane molecule—thus named for being cube-shaped. He was smart, he was cultured, he was a delight to interact with. To meet him was in itself a treat and well worth my hard work while at the University of Chicago.

From Chicago, I went to Ithaca, in the State of New York. Not only am I fond of the chemistry department at Cornell, I love Ithaca, the small university town nested in Upstate New York in the Finger Lakes region. It is an area of rolling hills, also of heavy wintery snowfall from a lakeside effect. When you arrive at Cornell for the first time, the advice given you is LAYERS, i.e., stay warm by accumulating layers of clothing. My stays at Princeton during the Sixties sold me on small university towns—not only the intellects of the US, this meritocratic country, but vastly more importantly, its soul—and Ithaca is definitely one, even more so than Princeton. I thus vastly enjoyed several stays there. I shall focus here on one only, that of 1986. I had rented from the Newtons an isolated house on Connecticut Hill, about 10 miles from the campus and downtown Ithaca. Relatively often, Valerie and I would be driving home and Carl Sagan would pass us in his conspicuous Porsche, with the PHOBOS license plate. Another star on the faculty, at least to me, was the physicist Hans Bethe (1906–2005).

My various stays at Cornell were engineered by Roald Hoffmann (b 1937) who graced me with his friendship. We wrote together a few essays. At lunchtime, he introduced me in the Faculty Club to some of his friends; from various departments, such as: Alain Seznec (1930–2017) and David I. Grossvogel (1925–2017), in French; Gordon M. Messing (1917–2002), from Classics and Linguistics; Isadore (Stuart) Blumen (1914–1991) in History; Anil Nerode (b 1932) in maths—he still teaches at the time of writing; and Vinay Ambegaokar (b 1934) in physics: a most distinguished group of scholars, however, in 1986, most regretfully I was far too busy to seek them out.

To some extent, because I had other acquaintances among Cornell academics. There were these outstanding pioneers of communication chemistry, Tom (Eisner) (1929–2011) and Jerry (Meinwald) (1927–2018). In 1986, Tom explained lumi-nously to me the evolutionary success of insects as a group: they are discontinuous eaters. Nematodes, in contrast, are continuous eaters. This is what led insects to develop a social organization. I met also with Jerry and his wife Charlotte. They were close enough friends to have visited my mother in Grenoble on one of their trips and had played chamber music together with her—Jerry was a very good flutist. In February 1986, they proudly showed me their 2-month old baby.

Other friends among the chemistry faculty included Simon H. Bauer (1911–2013), who remained active very late in life, the astringent David B. (Dave) Collum, whom it was always a pleasure to talk to, Bruce Ganem (b 1948) and Fred McLafferty (b 1923), Charles F. (Charlie) Wilcox (b 1930), Barry K. Carpenter (b 1949), etc. In addition, I had the treat of colleagues in other departments than chemistry. I have mentioned already in this book Mike Abrams (1912–2015), who had made Cornell safe—so to say—for Vladimir Nabokov (1899–1977).

Treat through lectures. To me, academia is uplifting—in like manner to a thermal making a glider soar. Two persons especially come to mind, during my 1986 stays in Ithaca. Carlos Fuentes (1928–2012), the Mexican novelist, was one. He presented a lecture series. He sought out Roald Hoffmann. And Roald invited me to join him for several dinners together with the Fuentes. They were festivals of culture.

Plus, there were the Fuentes lectures—extensively prepared, constantly informa-tive and a delight to listen to. Take for instance the lecture given on February 24th: looking at my notes, this master of oratory went through quotations at great speed which gave the well-founded impression of a vast culture and deep familiarity with their authors. His delivery put emphasis on sibilants and dentals, which greatly helped auditory reception and comprehension. Another asset of his presentations was what I can only term recitatives, by analogy with operas and cantatas of the Baroque: regular outbreaks of rhetoric that gave rhythm and pace to the whole flow.

Fuentes talked, among other topics, about the discovery of the New World; about its repercussions for Shakespeare and Cervantes. The Spanish discoverers were intent, not on reality but on fantasy: to them the Americas were a hyperbolic nature, disproportionate, incommensurate, to marvel at. To Cervantes, it was a servant of diversity, communicating mutation in mankind's idea of the universe. Whereas to Shakespeare, the New World shattered the space of humanism and vastly expanded the universe.

Fuentes was visiting, from Harvard at that time. I attended also regular classes by a Cornell professor of philosophy, Gregory Vlastos (1907–1991). He had a single subject, Socrates—and he was just amazing about the Greek philosopher, his trial and death (as Nietzsche emphasized, these and the parallel case of Jesus, some centuries later, were founding dramas for Western thought and religion). Vlastos talked about Socrates as if he were a personal acquaintance of his, the familiarity was stupendous—very moving. Not so much about Socrates the historical person; but about Socrates as a fiction, as a rhetorical device by Plato, as a mouthpiece for his own ideas. Vlastos thus contrasted two Socrates, that of the early period and that of the middle period, in the *Dialogues* of Plato. The former presented a boldly assertive metaphysical system: all knowledge is innate and souls are transmigrating. The latter, by contrast, was metaphysically reticent with respect to properties of the soul. To listen to a class by Vlastos was a rich experience, as moving emotionally as it was perceptive intellectually.

Given my happiness in Ithaca and at Cornell, did I do anything to make it more permanent? The answer is a qualified yes. I hinted or said as much to the then chairman in Chemistry, John Wiesenfeld (b 1944), to my friend Roald and to another good friend, Earl Peters (1927–2017), the Executive Director of the Department. For what kind of a position? A combination of overseeing the nmr facility (one-third of my time) and of running a research group, focused on chemistry with clay auxiliaries and catalysts (two-thirds of the time). The nmr instruments indeed had a director, David Rice, who in subsequent decades made a brilliant career at Cornell— graduating to become director of another high-tech facility, the Cornell Electron Storage Ring. In 1986, David Rice had been at Cornell for 1 year only. At the nmr lab, he supervised an electronics technician, David Fuller.

There were complications though, both personal and professional. My marriage to Martine was coming to an end. Indeed, I was spending these months in Ithaca together with Valerie, my future wife. The professional issues were at least equally complex. How to deal with my continued appointment at the University of Liège, in Belgium. Starting my new appointment, at the *École polytechnique*, in Palaiseau, near Paris.

Without these issues being resolved, I was called upon to present a seminar in the Chemistry Department. At least a dozen faculty members attended, which meant that my appointment as a professor at Cornell was being considered. Did it go any further, I never found out.

What killed it, I believe, were two factors. Maurice Bernard (b 1928), the then head at *École polytechnique* (1983–1990), visited Ithaca at that time. I believe that he talked Roald Hoffmann into not interfering with my start at the *École*. And I became intensely proud of my returning to France in such a prestigious position—prestigious but fraught with insoluble problems. My vanity thus made me oblivious of the dangers on the road ahead, that would bring my career to a premature end, in 1999.

I accepted other visiting professorships in addition to the preceding. To give them brief mentions, there was Johns Hopkins, in Baltimore (1981). It was shared by the two Departments, French Literature and Chemistry. Since I enjoyed use of offices in both, plus a carrel in the Milton Eisenhower Library, people had difficulty locating my whereabouts at any time and I had total peace to do my work. Chemistry-wise, the late Gary Posner (1943–2018) was a colleague whom I greatly esteemed and admired for the quality and originality of his work.

In September 1985, I served another visiting professorship, in Okazaki, Japan, at the Institute of Molecular Science. I was hosted in the Department of Applied Molecular Science, headed by Professor Iwamura Hiizu (b 1934). I enjoyed very much my interaction and discussions with a fellow guest professor, Gerhard L. Closs (1928–1992), from the University of Chicago. From Okazaki, I took a number of side trips, such as to Pearl Island, Nakasone and on a stretch of the Old Tokaido Highway, in the mountains reached by train from Nagoya.

In 1990 or 1991, I was hosted at Berkeley by Professor Gabor Somorjai (b 1935), in his Institute on top of the campus, in building 66. We had first met at a Bürgenstock Stereochemistry conference, when he had invited me for a 1-month stay in his Institute of surface chemistry and catalysis. On the vast Berkeley campus, I loved walking by the eucalyptus trees: I would collect their fruit, for their indentations—from threefold to sixfold symmetries, most often.

During the summer of 1992, Valerie and I, together with Aline my youngest child, stayed for 1 month at the Brookhaven National Laboratory. My invitation came from Leon Petrakis. He then headed the Division of Applied Science. He meant for me to assist him with preparing a report for publication—which I duly complied with.

I took advantage of my being on Long Island to convene a reunion of my former research group in Princeton. This is a nice and long-lasting recollection, of spending several hours on a Sunday with those friends, all former graduate students whom I had supervised: Arnie and Myrna Speert—Bill and Elaine Frankle—Sarangan—did his wife accompany him, I do not recall—Art Greenberg.

To close this chapter, my visiting professorships were means for moves both outwards and inwards. Outwards, to sympathize and interact with fellow-academics—nowhere was this more pronounced than at Cornell, during my several stays in Ithaca, that wonderful place. Inwards, as I was assumed to be up to the challenge of performing at top capacity in a new environment—self-reliance, to name it properly.

Skipping the Norms

17

My entire career was spent in academia. I was part of the university system in three countries, France, the United States and Belgium. In addition, visiting professorships gave me an insider view at those of Germany and Switzerland. This chapter will chronicle some of the changes that took place during the second half of the twentieth century, from patronage to entrepreneurship, from gentlemanly pursuit to rat race and from letters of recommendation to scientometrics and rankings.

My very beginnings took place at the Sorbonne in Paris and at the nearby new university in Orsay. I started as a teaching assistant in biochemistry for Edgar Lederer (1908–1988), a professor at the Sorbonne who moved to Orsay. After he failed to convince me to enter CNRS, I became an instructor at Orsay first in electronics (!) and afterwards in general chemistry. In turning down his offer of a position in CNRS, that extremely French organization; I was intuitively motivated by the alliance of teaching and research. I realized later the wisdom of that choice, teaching and research feed one another.

After our return from Princeton to Paris, in order for Martine to complete her medical studies, I worked at the *Institut de chimie des substances naturelles*, in Gif-sur-Yvette, south of Paris. It had two parts, directed each by Maurice-Marie Janot and Edgar Lederer. He was nominally my supervisor. I collaborated with some of his coworkers.

Substances naturelles is in French what natural products stand for in English. In the Sixties, it was a subfield of organic chemistry, with a distinguished past and a highly promising future, that already pointed the way toward some form of unification of chemistry and biology.

To a significant extent, the rich past came from the association with a French luxury industry, the manufacture of perfumes. It had arguably reached its pinnacle about 1900 and the Paris World Exhibition.

In the Sixties, there was still a rather flourishing manufacturing of perfumes in Grasse, a little town next to the Riviera above Nice. They were made from both locally grown flowers and imported ingredients, also from natural sources. The best known and then still existing factory was Roure Bertrand—it closed in the Nineties.

In the Sixties, when it had already been purchased by Hoffmann-La Roche from Basel, it was still very active =: on a yearly basis, large volumes of flowers were transformed into essential oils, ca. 100 tons of jasmine, an equivalent tonnage of mimosa, 75 tons of roses, etc. Roure-Bertrand were suppliers to designers of perfumes—known as *nez* (noses) in French—some of which had attained the status of classics: Opium and Kouros (Yves Saint Laurent), Loulou (Cacharel), Poison (Christian Dior), etc.

Why did it close towards the end of the century? Because of competition from other countries, where there was less emphasis on natural sources and where synthetic molecules coexisted with natural products as ingredients of perfumes. There was a large American corporation, International Flavors and Fragrances, in New Jersey. And there was, in Switzerland, both Firmenich and Givaudan—Professor Lederer was a consultant to Firmenich—Givaudan would merge with Roure during the Nineties.

My work, to return to my collaborations with various Lederer coworkers, was on the area of diterpenes mostly, i.e., natural products whose molecules have 20 carbon atoms, resulting from the coming together of four C_5 isoprene units. The number of known diterpenes is very large, they come in very diverse chemical structures—I helped to elucidate a few (Fig. 17.1).

During the first half of the twentieth century, some of the leading French organic chemists were active at the interface of perfume chemistry, often involved for that reason with aldehydes, a class of highly volatile molecules. There was Georges Darzens (1867–1954), one of my predecessors at the *École polytechnique*. At the turn of the century, he directed the R&D laboratory in Paris of L. T. Piver, a major and leading manufacturer of perfumes. There was Léonce Bert (1892–1972), in Clermont-Ferrand, whose epic and tragic story I shall present in these pages. There was Georges Dupont (1884–1958) at the *École normale supérieure* and his student Guy Ourisson (1926–2006), who would go on to the University of Strasbourg. and whom we shall meet again in these pages.

My departure from Orsay and the French academic landscape and back to Princeton, where I had already spent a postdoctoral year in 1962–1963, needs retelling: it is picturesque, it shines also a light on a French blind spot, call it a lack of practical sense within both myself and the universities. This happened in the spring of 1966. I was an instructor in Orsay, running the chemistry labs for premeds. One day, during lunchbreak—sacrosanct in France as is well-known—a huge truck arrived. It was loaded with huge wooden cabinets to equip that new lab. The truckdriver said the unloading was not his task. Hence, both Georges Bram (1937–2004), the other instructor and I carried the cabinets into the building. Upon return to our apartment that evening, hurting all over, I went straight to bed. My wife Martine said, "you have not gone through such extensive study to be a handiman. Let's return to the US. Do write asap to universities there, asking for a teaching position." Which I did. I applied to MIT, the University of Oregon and the University of Wyoming. Why Wyoming? Because as a youngster I had been thrilled by the Flicka books! And I asked Paul Schleyer at Princeton to send letters of recommendation to these institutions. Paul failed to comply but just a couple of

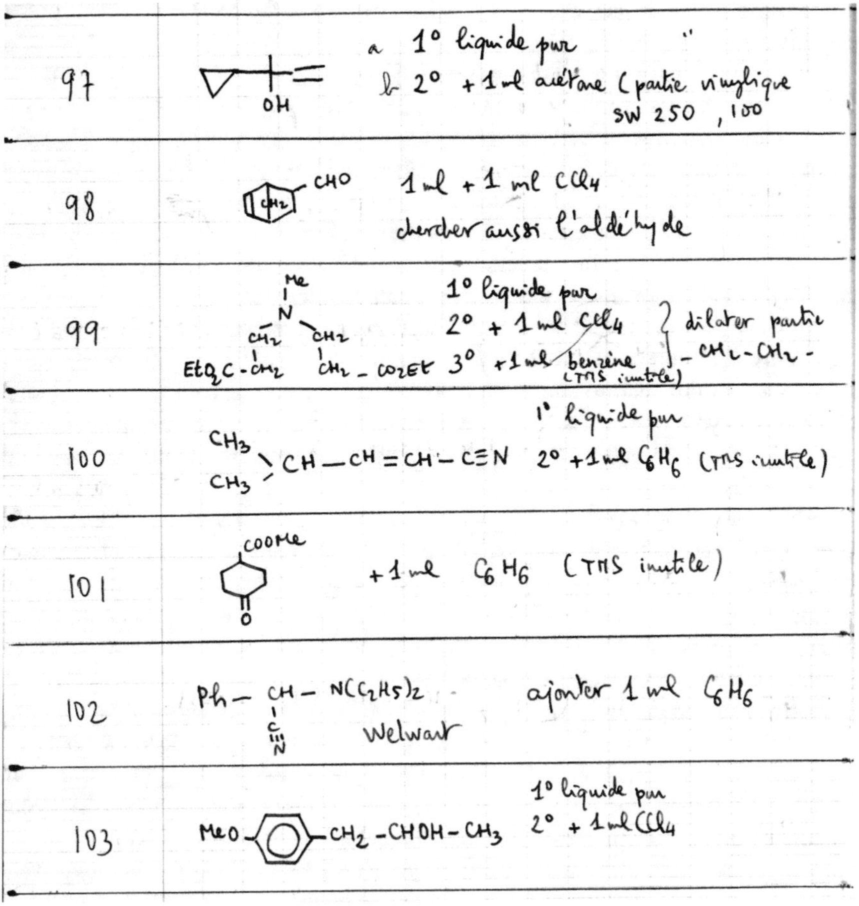

Fig. 17.1 Nmr spectra taken on Friday June 15 1962. A page from my notebook

weeks later I received a letter from the President of Princeton appointing me an assistant professor in chemistry.

This was representative of patronage, as it ruled at the time American colleges and universities—at least those with the top rankings. Within academics, prestige and power went hand in hand. Prestige accrued from the quality of one's work—and from the prestige of the institution you were associated with. Power allowed one to give pupils and coworkers the best available positions. A couple of examples in support: Robert B. Woodward (1917–1979), genius and Nobel prizewinner, made certain with phone calls that his coworkers got the best available jobs, in either academia or industry. Paul D. Bartlett (1907–1997), who also taught at Harvard and was a pioneer in physical organic chemistry, placed many of his former students in important academic positions (Paul Schleyer was one of them). Howard E. Zimmerman (1926–2012) at the University of Wisconsin even proudly displayed in his office a world map with little flags pinned upon the locations of his former coworkers.

Science as I have known it, having started my career in America, is meritocratic. The best persons rise to the top—that was the ideology. Unduly naive? Probably. The organization was indeed hierarchical and a majority among colleagues were keenly aware of rankings. Ranking of universities. Ranking of departments within a university. The implicit but nevertheless powerful ranking within a department.

For instance, at Princeton in the Sixties, the department of chemistry ranked about 12th in the US. Within the university, it ranked below mathematics, physics and philosophy—all three among the very best nationwide.

But what were the bases for such rankings? Faculty members, predominantly, and their individual outputs—publications primarily. Yes, publish or perish applied. I shall return to this dictum and its consequences. But first a word about the moral contract with your coworkers, graduate students foremost. They give you their work, often also in my experience their respect and affection. You owe them in return a lift up the academic ladder.

Which brings up a rule-of-thumb I have evolved from my experience over the years. It is totally pragmatic and grossly unjust in its brutal cruelty. It is the rule of one-in-four. Out of four undergraduate doing research in your lab, one is good enough for graduate study. Out of four graduate students, one is good enough for postdoctoral status. Out of four postdocs, one may be good enough to be recommended for a faculty position, somewhere.

But what are the quality criteria for scientists, their appointments and promotion? During the period of my career, these were in no particular order: the school you went to as an undergraduate—that of your graduate study—whom you chose as an adviser—the papers you published and the journals in which they appeared—your lecturing skills—your prowess as a communicator—your ability as a fund raiser—the perceived status of your work area—your originality. Yes, the last two were contradictory: as in any human group, there is pressure to conform to the norms of the group; but there is also grudging respect for the individual transgressor—I was one (see the title of this chapter).

Coming back to university appointments, what I witnessed in both Belgium and France, where excellence was not at a premium, during the Sixties and Seventies predominantly, was mere number of papers published, somewhat irrespective of the quality of the attendant journals. Yes, a shocking procedure—but one that allowed the full impact of patronage. In the US during the same period was dominance by letters of recommendation. From whom? Influential people in positions of power. Indeed, such reliance upon the all-important group of peers goes together with peer-review in upholding quality standards in scientific life. The downside of basing academic appointments on letters of recommendation is all-too-obvious: individuals kowtowing to figures of authority in the—unfortunately realistic hope—their sub-servience will be rewarded with a favor: "it is not what you know, it is whom you know," as the saying goes.

And what about more objective criteria? I did witness during the Sixties the birth and rise of scientometrics, pioneered by Eugene Garfield (1925–2017). I subscribed for many years to *Current Contents Physical Sciences*, an extremely useful service he introduced. I watched the index he would regularly publish on the relative impact of the various journals, led by *JACS* in which I was (fortunate or lucky) enough to publish quite a few of my papers.

A simplistic point often made about the academic job market is "that a professor, in the US say, had umpteen graduate students during his/her career, only one of which will be a professor in turn." Even though the assertion has the appearance of logic, the empirical evidence fails to support it.

Take my example, I have had *about a dozen successors*, viz. Arnold Speert, a professor and president of William Paterson College of New Jersey; Christian Detellier (b 1949), a professor and vice-president at the University of Ottawa; three professors at the University of Liège, viz. André Cornélis (1941–2021), Jean Grandjean (b 1945) and Lionel Delaude; Arthur Greenberg (b 1946), a professor and a Dean at the University of New Hampshire; Wacław Kołodziejski (b 1949), a professor at the Warsaw University of Medicine; Peter L. Rinaldi, a professor at the University of Akron; and I should not fail to mention also the late Janusz Baran, University of Szczecin and Haggai Gilboa, of the Technion in Haifa.

The so-called academic job market goes through periods of expansion and contraction. Expansion: take again my example. When I started as a faculty member at Princeton, in 1966, there were about a dozen assistant professors hired simultaneously in the chemistry department. This was a consequence of Sputnik and the Cold War, the United States was then making a huge effort to expand science education and research. Contraction: as just an example, also from the US, the abolishment in 1994 of a mandatory retirement age as discriminatory. Expansion and contraction also go through cycles, for instance expansion in the Sixties was followed by contraction in the Seventies due to the economic recession of the mid-1970s.

Was the French system different at that time of the Sixties? Yes and no.

It was based on a combination of *mandarinat* and localism. The latter first.

French academic appointments are fraught with a national disease, that of localism: rather systematically, a local candidate gets the preference. Since this may sound strange, it needs an attempt at explanation.

Perhaps the best way to look at it is from the perspective of the *terroir*, i.e., the notion of an outstanding wine, cheese or landscape, originating in a given small French region. President De Gaulle, upon being asked if it was difficult to govern France answered "what do you think, the country sports 350 different cheeses." Why do the French root themselves in a small district rather than in the country as a whole? It can be understood as a relic of history: the distant religion wars in the sixteenth century that tore the country apart; and more recently World War II, the German Occupation and the split between collaborators and Resistance fighters. Such turmoil leaves durable traces.

I have direct knowledge of two episodes in which indeed a local candidate to a university position in France—rather than the best person in the country—got the nod. During the late 1960s, when I was on the faculty at Princeton, two colleagues in Grenoble, Didier Gagnaire (ca 1930–2014) and André Rassat (1932–2005) pushed me to introduce my candidacy to a professorship at the University of Grenoble. They told me that my chances were nil and that a local candidate named Jean-Louis Pierre (b 1942?) most probably would be chosen. I agreed to become a candidate, since Grenoble was my hometown and both Didier and André were esteemed friends. It failed as predicted. They were told my candidacy documents had not arrived, a lie.

The second episode dates to the mid-1970s when I was on the faculty at the University of Liège. A colleague from the French university of Lille, Claude Loucheux (1930–2012), who at the time chaired the department of chemistry there, asked both myself and Heinz G. Viehe (1929–2010) who taught at the University of Louvain-la-Neuve, to serve as foreign observers: they were about to replace a retiring professor, Charles Glacet (1911–1986) in synthetic organic chemistry. Loucheux had adequately publicized the vacancy and there were half-a-dozen well-qualified external candidates, from the Paris area—two or three even outstanding. One name I remember is Max Malacria (b. 1949). An inferior local candidate, Daniel Couturier, was picked instead.

In France, *mandarinat* was the name of an institution that for being unofficial was all the more effective. Based on perceived seniority and a university ranking with the Parisian Sorbonne at the top, it was the naked use if power by the professors imbued with it—the so-called *mandarins*. This was nowhere more evident than in academic appointments. In feudal manner, these went to the pupils of these *mandarins*.

Who were they? In organic chemistry in the 60s, the Sorbonne professors were Charles Prévost (1899–1983) and Albert Kirrmann (1900–1974). Both were alumni from *École normale supérieure* which ensured their top level appointments. Unfortunately for French science, Prévost if bright was a lazy person. Kirrmann had spent the war years a prisoner at Buchenwald and came back a broken man. In any case, both were able to place their former coworkers in the (perceived) best positions within the French university system (Fig. 17.2).

There were also *mandarins* in provincial universities. An example, in organic chemistry also, was Jacques Metzger (1921–2014), the absolute king in Marseille. Was *mandarinat* deterimental to French science? Absolutely. I'll give a single example, that of Georges Champetier (1905–1980) in polymer chemistry. He blocked any appointment of Charles Sadron (1902–1993) in Paris, whether at the Sorbonne or at the *Collège de France*. From having been a postdoc at Caltech, Sadron had a vision for transforming French science. But Champetier would not tolerate any rivalry, polymers were his preserve in Paris.

As luck has it—my persistent luck is an undercurrent in this memoir—I witnessed *mandarinat* together with another scourge of French science, the influence of unionized scientists from SNESUP, at close quarters. At the end of the 1970s, when a professor in Liège—but still on leave as a *maître-assistant* (instructor) from the French university system—I was appointed to *Conseil supérieur des corps universitaires* (CSCU), in its chemistry section. This body oversaw professorial appointments for the entire country. The meetings, once or twice a year, brought together most of its members, between 15 and 20. To content myself with two recollections, the SNESUP representatives lobbied successfully for awarding a position in the southern provinces to one of their members with only a paltry record.

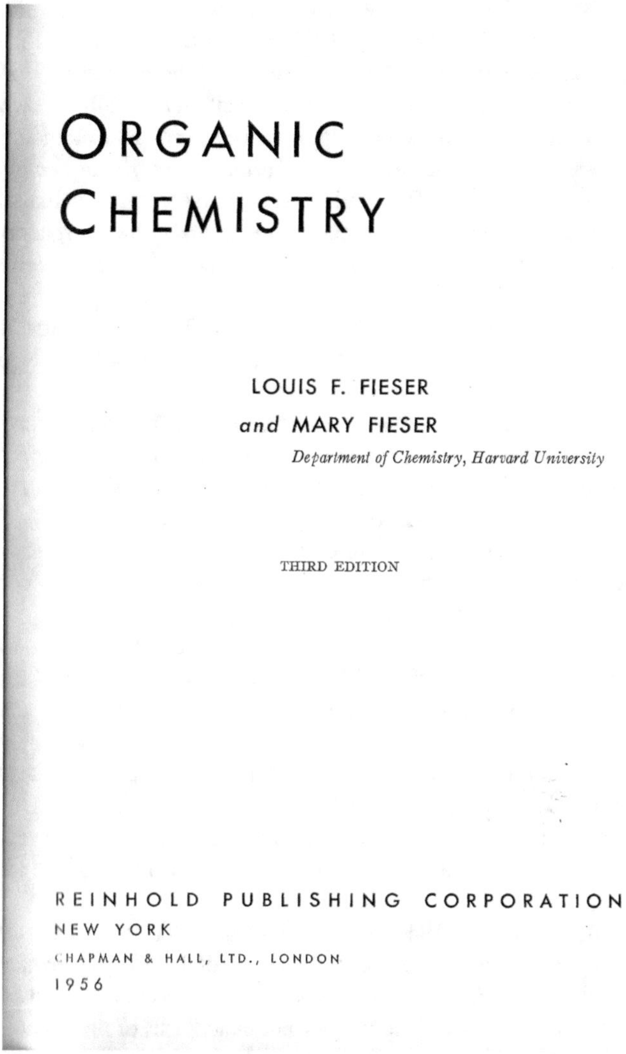

ORGANIC
CHEMISTRY

LOUIS F. FIESER

and MARY FIESER

Department of Chemistry, Harvard University

THIRD EDITION

REINHOLD PUBLISHING CORPORATION
NEW YORK
CHAPMAN & HALL, LTD., LONDON
1956

Fig. 17.2 Title page in my copy of *Organic Chemistry* by Fieser & Fieser, its reading in the early 60s by French organic chemists was semi-clandestine—one might term it samizdat

And there was among us Jacques-Émile Dubois (1920–2005), a professor with an admirable personality but also a talent for exercising absolute power and achieving his goals whatever opposition they met. I have yet to meet a more persistent lobbyist!

In short, I find the notion of an academic French job market ludicrous. What exists instead, or existed as I have known it, is cronyism. It can be blamed for its all-too-obvious shortcomings. It can also be praised for its virtues. Virtues of cronyism, of such blatant unegalitarian and undemocratic a selection process! Yes: those likewise of any meritocracy, based primarily on authentic achievements.

However, the shortcomings of cronyism eclipse its merits. I write this from experience, that of a privileged observer of the French CNRS—short for *Centre national de la recherche scientifique*. CNRS was founded, shortly before World War II, to boost scientific research in France. It provided positions for full-time work as a scientist, with no teaching strings attached. Stages in a career and the attendant salary followed upward the sequence: *chargé de recherches—maître de recherches—directeur de recherches*. Each CNRS worker was overseen and evaluated, on a yearly basis, by a commission made of specialists in the discipline—organometallic chemistry, say. Such organization amounted to a recipe for either excellence or mediocrity.

The former explained the world-famous school of French mathematics. The relevant commissions followed the precept by André Weil, appoint only persons better than you are.

The latter, as I have witnessed it, was chemistry. The main reason, cronyism. It was endemic among the leaders, from the tradition of patronage. It ruled the participants in the commissions from unions of scientists, such as SNESUP. I vividly recall Henri-Edouard Audier (1940–2016) the influential leader of SNCS, *Syndicat national des chercheurs scientifiques*, telling me "research is a job like any other." He meant a 9-to-5 job, with a couple months of summer vacation—to me the antithesis of the pursuit of knowledge for its own sake.

It is my contention that, following upon victory in the race to the Moon, the Cold War mentality in the US, starting with the Federal government, shifted none too subtly from improving the scientific education of the population and pure science in the mushrooming research universities, to reaping benefits from applied science and technology—follow the money. I witnessed thus during my career a definite shift—a shift, not a switch—from academic excellence to entrepreneurship. University appointments thus swerved from a regime of patronage to one of entrepreneurship—from paternalism to self-advancement, to put it bluntly. I have some examples in mind.

The first is George C. Levy (b 1944). A student of Saul Winstein (1912–1969) at UCLA, he was "orphaned" upon the untimely death of his all-powerful mentor. Thus, he did not continue in the line of his 1968 doctoral dissertation and switched instead to nmr and applications of computers to automated uses of nmr in chemistry. After having been hired in 1981 as a professor at the University of Syracuse, he built a small empire in applied science. He created a start-up company there, New Methods Research, Inc., to develop and market NMR data processing and NMR laboratory networking software based on general purpose computers and workstations.

My second example of an American academic who also shone with his entrepreneurship is Edward T. Samulski (b). A graduate student in polymer science at Princeton (1965–1969) with Arthur V. Tobolsky (1919–1972), working on liquid crystal polymers, he held academic appointments at the University of Connecticut (1972–1987) and since 1988 at the University of North Carolina. He has founded or co-founded four companies in succession: in 2004 Liquidia Technologies, set in Research Triangle NC, a company designing and producing drug particles uniformly and in a wide variety of compositions, size and shapes, for improved therapeutics. In 2008, Allotropica Technologies for producing high performance liquid crystal thermosets. In 2013, Carbon based in Redwood City CA and in Chapel Hill NC, producing polymers for three-dimensional printing. And most recently, in 2020, Blue Sky Polymers, also based in Chapel Hill NC to make novel high performance polymers.

My third case is of a colleague who opted out of the academic competition (grants and articles) and espoused instead an entrepreneurship of his own. This was John E. McMurry (b 1942). He had risen to the top of the profession, was a professor at Cornell and he had even devised a new reaction that bore his name. He also wrote a successful textbook. It enabled him to "drop out of the rat race" (his words), at the end of the century: he left Ithaca for early retirement in Oregon, where he worked from his home and continued writing successful chemistry textbooks, five in total. Two have had eight successive editions already. This decision of his to drop out is a telling sign of the overwhelming pressure on an academic in the US.

I'll return now to the issue of patronage, as it dominated academic appointments in the previous century. This is to mention candidacies of mine that did not fly high enough to succeed. In 1978, when a professor in Liège, a colleague I highly esteemed, Aksel A. Bothner-By (1921–2017), with an influential voice at Carnegie-Mellon University in Pittsburgh contacted me, my name had been mentioned for the chairmanship of the chemistry department there. I visited and gave a seminar but nothing came of it.

In 1991, I was a professor at *École polytechnique*. Maurice Bernard (b 1928), who had hired me in 1986 had been forced out of the directorship in 1990. Henceforth I would be devoid of a bulwark against attacks—that indeed occurred. I learned of a vacancy at the University of Neuchâtel in Switzerland. I applied, was invited for a seminar and was greeted most courteously. It amounted to nothing: a few weeks later, my good friend Pierre Potier (1934–2006) told me he had had a talk on the phone with Albert Eschenmoser (b 1925), from the Swiss Polytechnic in Zürich, who had told him he considered Neuchâtel part of his preserve. Positions there were reserved for his coworkers—which I believe indeed happened.

My unstable status at the *École polytechnique* encouraged me to pursue offers of a position elsewhere. One such came my way in 1992. Victor Hruby (b 1938), a professor at the University of Arizona in Tucson, informed me my name had been mentioned for the chairmanship of the Chemistry Department. When I answered the invitation and got there, I was informed that in the meanwhile other plans had been made.

Another offer came in 1993, from Tallahassee in Florida. Florida State University there had succeeded in acquiring a highly visible facility, the National High Magnetic Field Laboratory—in the process even prevailing over MIT. They were hiring three high-profile professors to run the attendant research. I was short-listed for one of these positions—but in number 2 position only. I travelled to Tallahassee and gave a talk. Geoffrey Bodenhausen (b 1951) was number 1 in the short-list and received that call. However, the all-powerful figure who had spearheaded the highly political application by FSU for this facility was a flamboyant autocrat.

Bodenhausen lasted only a year or two. Fortunately for him, the *École normale supérieure* in Paris gave him a position in 1996.

I am a rather private person who shies away from groups. "Far from the madding crowd" could be my motto. Another favorite of mine is by Henri Michaux, "he who sings in a group will put his brother in jail, when told to do so." A third quotation is more than apt, from Feynman's wife, "why do you care what other people think?"

And yet I am a scientist: is not science a collective advance fueled by competition? My career displays the converse: I have followed my nose wherever it led me without caring for the pressure to conform and ignoring the various fads I witnessed from a distance. My instinct—how to call it otherwise—thus shielded me from the dark side of scientific competition, others have abundantly documented it. To me science is primarily an intellectual pursuit.

I have seldom applied for research grants. The functioning of my research groups rested on support from the department, when at Princeton, and from the university, both in Liège and at the *École polytechnique*. It depended very much on the allocation of teaching assistantships, which allowed me to employ coworkers who carried out the actual research work at the bench.

And I was consistently lucky. Which I can illustrate with two episodes dating from my time in Liège. About midway, I applied to FNRS (*Fonds national de la recherche scientifique*) for a high-field 300 MHz nmr spectrometer. Unbeknownst to me, all my colleagues in the *Institut de chimie* did likewise. FNRS people were in a bind. They decided to allocate the instrument to the University of Liège, leaving it to its administration to arbitrate between the two rival requests. The then President of the University, Emile H. Betz (1919–2012), made a heroic decision: he chose to give me the instrument thus denying it to the others. Needless to say, it was resented.

My other piece of luck was to obtain around the turn of the Seventies into the Eighties a most generous funding allocation from the Belgian *Programmation de la politique scientifique*, a so-called *Action concertée*: for about 5 years it shielded my laboratory from funding worries.

I close this chapter on a note of serenity. The best academic locale I was blessed with during my career, to me haven and heaven both, was the chemistry department at Cornell. What was so special about it? In brief, at the times I served a visiting professorship there—in the late 80s and again in 2000 and 2001, after my formal retirement from my European appointments—was that it was a community of equals. Its assets were life in Ithaca, in Upstate New York with its rolling hills and the Finger Lakes, a remote, isolated, small university town. The harsh winter climate with abundant snowfalls imposed solidarity. Conversely, the world came to Ithaca, in the form of visiting lecturers of the first rank. Those I recall most vividly were the bishop Desmond Tutu (b 1931) from South Africa and the Mexican novelist Carlos Fuentes (1928–2012). The chemistry department had no pecking order. There were no tangible tensions or rivalry between colleagues. I loved the club-like atmosphere of the departmental meetings, where I was treated as an honored guest. They had a rotating chairmanship, so that no faculty member would stay too long in power. There was a Faculty Club for meeting at lunchtime colleagues from other departments. And a world-class library. I noted that my fellow professors on the whole stayed many years at Cornell, often for the duration of their whole career. To illustrate what I felt during my stays there I can only compare it to the Abbey of Thélème, the humanist utopian community of scholars described by Rabelais in the sixteenth century—the inscription on the gate to that abbey prescribed who was unwelcome: "hypocrites, bigots, the pox-ridden, Goths, Magoths, straw-chewing law clerks, usurious grinches, old or officious judges, and burners of heretics."

Meaningless Fads, Meaningful Changes   **18**

Change or rather changes. Taken in isolation, a change may carry little meaning. Taken together, a few changes may point to a deeper significance, a trend or perhaps even a shift.

My career as a chemist lasted long enough for changes, big and small. I'll start with the most mundane. In the early 60s, most chemists worked at the bench and they wore a lab coat. In the 70s, after the pivotal year 1968, jeans became the new uniform in the profession. By the 80s, there had been a mass migration to small desks in front of a computer screen.

Such changes ought to be taken seriously. If I put them together with a story such as that, told in an earlier chapter, of the highly successful Professor Hogeveen (1935–2015) dropping out of chemistry and becoming an artist. Or the episode of at first vilification by the chemical industry of F. Sherwood Rowland (1927–2012) and Mario J. Molina (1943–2020) for their discovery of—to put it briefly—the ozone hole caused by chlorofluorocarbons, with the latter turnaround and the Montreal Protocol (1987). Or the vogue nowadays of biofoods and farming: an irresistible take-home lesson is a desire by chemists to feel more a part of society at large and to shoulder less of a collective guilt for the fallout from harmful chemicals of all kinds, starting with DDT. Rachel Carson's (1907–1964) *Silent Spring* (1962) did to chemists in 1962 what Hiroshima did to physicists in 1945: it made us ashamed.

Coming back to daily laboratory activities, what about work with a vacuum line? I saw its introduction in the 60s from organometallics, it had become ubiquitous by the mid or late 70s. The glove box was its companion tool. I do not need to belabor the then novel emphasis on safety of workers, not only of air-sensitive materials.

Turning from the laboratory to its final product, the publication, what is the meaning of a flamboyant rhetorical device such as—starting in the 90s—mention of the Holy Grail in such and such subspecialty? Assuredly, an internalized weakness that this hyperbole is meant to hide.

My first combination of recollections and analysis concerns computational chemistry: in what way has it changed chemistry? Is it too early to try and evaluate costs

P. Laszlo, *A Life and Career in Chemistry*,
https://doi.org/10.1007/978-3-030-82393-1_18

and benefits? My main impression, very positive, is that it has strengthened continuity and cumulativeness in the science. Through simulations it allows chemists to take a close look at weak interactions, such as intermolecular forces responsible for key biological events, such as the docking of a ligand in a receptor site—a protein typically.

Computations would allow me today to look at the interaction of camphor with benzene solvent molecules—the 1971 paper with Ed Engler referred to in an earlier chapter—from a different angle: finding the geometries for camphor interacting at first with two benzene molecules, then four, eight, 16 … Once these calculated situations ascertained, I would run a molecular dynamics program to provide a picture of the time evolution. I feel confident that, in this manner, our earlier conclusions would not only be bolstered—a trivial satisfaction—but also extended.

Computational chemistry has already made a deep mark for half-a-century. One might term it instead, computational analysis, by analogy with conformational analysis—a tool more briefly during the Sixties. As is well known, computational chemistry is a hybrid set of methodologies combining quantum chemical and molecular mechanical calculations. Several Nobel prizes were awarded for its development. These prizewinners that I was most impressed by their brilliance were John A. Pople (1925–2004) and Martin Karplus (b 1930). Other names deserve acknowledgment, most of all Norman L. Allinger (1928–2020) for his pioneering work in molecular mechanics—at the University of Georgia, which now hosts a Center of Computational Chemistry—and their contemporary, also my former mentor, colleague and friend Paul von Ragué Schleyer (1930–2014) who ended his career at the University of Georgia. In the same spirit of eulogy, I wish to mention QCPE—an acronym for the Quantum Chemistry Program Exchange—at the University of Indiana in Bloomington. Started in 1963, it fulfilled for years until the end of the century the essential function of providing available software at cost to any applicant.

I cannot help taking note at this point of a trend associated with computational chemistry. Sometimes and this is not a rare occurrence, articles in that area have but a single author. You can just imagine the gal or the guy in front of the computer screen running a piece of software and scrutinizing the results. I like it! Ample time to try and reverse the trend of papers showing multiple authors, all too often I regret to say, a small crowd.

In 2020, I am able to continue watching eagerly the development of biological chemistry. I see it without surprise, the beast is still the same it was in the Sixties or Seventies. The changes occurred mostly in the clothes it is wearing. An article that bore three or four signatures as a maximum, now can routinely carry a dozen or two—or even 88, in a 2011 paper in *Nature* on crystallography of proteins. Why such an inflation? Because biological chemistry has turned into Big Science: in the same way as physics articles from CERN and its hugely expensive instrumentation were collective and cooperative ventures enlisting a number of scientific teams, so does this scientific sector nowadays. However, for using hugely expensive tools such as synchrotron radiation, it has not changed significantly in depth or spirit.

One of the key issues remains the multi-tasking role of proteins, of which there are 20,000–25,000 non-redundant ones in organisms. Of human proteins, more than 90% had been detected by September 2020. In my time, I was glad to study a handful of calcium-binding proteins. Nowadays, research examines for instance, post-translational modifications critical to a protein's function, proteins responsible for the circadian clock in single cells or the ageing process in tissues with tools such as high-throughput analysis of biological processes, mass-spectrometric characterization and interaction with drug molecules. Proteins in the body have diverse functions. One may think first of enzymes, i.e., catalysts for the chemical reactions whose sum total make up metabolism. There are structural proteins, such as actin that ensures muscular contraction. Proteins are used also to transport and store useful chemicals, such as iron. Antibodies, immunoglobulins for instance, fight intrusion by foreign agents, such as bacteria. And hormone, growth hormone for instance, signals the onset or the ending of a biological process.

To return to the name chemical biology why not continue with the former term for that field, biochemistry? Obviously, one of the reasons is the need for novelty, each generation fancies itself as the lone bearer of truth, previous ones having been shrouded in the clouds of ignorance! A key reason perhaps is that chemical biology is more systems-oriented than biochemistry was.

During the active period of my career, we called it solid-state chemistry. Nowadays, it is termed materials science. Incorrectly, in my view, since such "science" is meant for applications much more than it is turned towards gaining fundamental knowledge.

What it shows and even advertises is the suffocating embrace of techniques on science. The term technoscience was devised and used since the Eighties by Bruno Latour (b 1947) and his followers, the self-named postmoderns. In their view, there was and is a continuum between science and technology. I submit that, while not being incorrect, this viewpoint—while hostile to the very notion of a capitalistic technoscience—was heavily influenced by consumerism and the values—or absence thereof—it embraced.

Capitalistic technoscience? Indeed, as it appears from a mere list of the proliferating journals that the American Chemical Society publishes in that area: *Chemistry of Materials—Materials Letters—Biomaterials—Accounts of Materials Research—Applied Biomaterials—Applied Electronic Materials—Applied Materials & Interfaces—Applied Nano Materials—Applied Polymer Materials* (no fewer than nine journals at the time of writing, the end of 2020). From just glancing at the subject matter and at the authors of the articles, three notions are evident: first and since novel technologies embody tomorrow's daily life, advertising and marketing tools apply already; how else should one account for the hype in some of the papers? Some of the authors blow their own horn way too stridently. Second, obviously China and its academic institutions invest heavily in this area. Thirdly, some authors have no qualms at reinventing the wheel and presenting as their inventions stuff that has been around for a while—a few generations sometimes.

Science is subject to fashions—nowadays termed megatrends by their enthusiastic supporters. Scientific fashions are very useful, bringing about a sharp distinction between leaders and followers, the few and the many, the prophets and the crowds.

During my career, I witnessed a few such fashions. There was the classical-nonclassical ion controversy that raged rather briefly, about the true structure of the 2-norbornyl cation. It brought about the demise of the unsuspecting physical organic chemistry, at the time—the Sixties—the leading subdiscipline in organic chemistry—perhaps even within the whole of chemistry.

I witnessed the polywater fashion, also relatively short-lived. It came about from an artefact, the ease at which water dissolves silica and turns it into a gel. Felix Franks (1926–2016) wrote a lovely little book (1981) on the history of that episode. He likened it to an epidemic, which indeed it was.

Another epidemic-like scientific fashion, short-lived though it happened to be, was the cold fusion craze. There were many other scientific fashions, all of which were indeed epidemic-like, all of which brought about the sharp sociological distinction between leaders and followers. They even extended into history of science, with the takeover of the field by sociologists who named this new area Science Studies.

Then came along the nano scientific fashion, during the Eighties. It was born from the near conjunction, time-wise, of the invention in 1981 of a technique, scanning tunneling microscopy, by Gerd Binnig (b 1947) and Heinrich Rohrer (1933–2013); followed by atomic force microscopy in 1985; and the discovery, also in 1985, of fullerene (C_{60}) by Robert Curl (b 1933), Harold Kroto (1939–2016) and Rich Smalley (1943–2005).

Just a word about the latter since Rich was among the graduate students I taught at Princeton. He followed my class in structural analysis by spectroscopic methods. He was older than other graduate students in his year and thus more mature, having worked several years in industry before deciding to return to school. Later on, when I visited Rice University in Houston and gave a seminar there, he was my host—most friendly and congenial. In addition, he was, not only a serious dedicated scientist, also a deep thinker—who died way too young.

As for the nano studies, I espouse Joachim Schummer's well-documented and nicely argued view of the intrinsic heterogeneity within that area. Nano is an umbrella-like term for a plethora of mostly applied pieces of work, aiming for remunerative technological applications and thus coined for marketing and advertising purposes.

A survey was made of the top 10 nanotechnology research organizations during the period 1970–2012, based upon the number of publications issuing from each such source. The top three were the Chinese Academy of Sciences, the Russian Academy of Sciences and the French CNRS. The following three were Japanese universities. Such a ranking testifies to a me-too mentality, that I find highly detrimental to the quality of science, it shows a pressure to conform where, even in applied science, the future belongs to the imaginative loners.

With the arrival of the twenty-first century, scientific journals underwent significant changes and not for the better. I shall comment here on proliferation of new specialized journals, concentration of scientific publishing, plundering of university libraries and predatory journals. Yes, these derogatory words reflect a deplorable reality.

During the second half of the twentieth century, the United States led in scientific research. Accordingly, it led also scholarly publishing with peer review. In my field of chemistry, journals published by the American Chemical Society (ACS) were dominant—starting with the already mentioned *JACS*. Other prominent ACS journals, in which I also published were *Accounts of Chemical Research*, the journals of *Chemical Education*, of *Organic Chemistry*, of *Physical Chemistry* and *Langmuir*. The total number of ACS journals was about 15.

Now, in 2020, the ACS publishes about 60 different journals. I find such inflation worrisome. For sure, there are some good reasons to explain and perhaps justify it. Arrival in the laboratories of new countries, China foremost, and new generations of scientists needs being taken into consideration—together with the exponential rise in numbers of scientists and of papers published. I recall a statistic from the Sixties: the average scientist published a single paper in his lifetime; clearly, this is no longer the case.

New specialized journals: I did mention already ACS journals in various subfields of materials science and of chemical biology. Their appearance came about at the confluence of a few factors: the regrettable in my view merging of science and technology, with the former becoming dominated by the latter, an epiphenomenon of ultraliberal capitalism; rivalry between a learned society such as ACS and commercial publishers (Elsevier, Wiley, Springer, …); ego trips of scientists eager to become appointed to the editorship of a journal, with the attendant inebriating feeling of power-wielding.

At this point, examples are necessary. *Tetrahedron Letters*, started in 1959 and already mentioned is a prime example of a monopoly exploited by a commercial publisher—Pergamon Press, the founder, later absorbed by Elsevier. It had become, during the four last decades of the twentieth century, a favored means of rapid communication among organic chemists, such as myself. However, the cost to libraries of its yearly subscription was obscene: in 2005 already, it was costing US$11,595 versus the somewhat more reasonable $3865 for the ACS-published *Organic Letters*. The ACS started publishing that journal in 1999 to provide an alternative to the grossly-overpriced *Tetrahedron Letters*—the annual institutional subscription of which is now (December 2020) priced at $21,621 versus $7148 for *Organic Letters*. This timely initiative shows the ACS as a benevolent association of scholars stepping in to moderate the excesses of commercial publishing. Likewise, with publishing by the Royal Society of Chemistry of *PhysChemComm* as an alternative on the Web to *Chemical Physics Letters*, offered by the same commercial publisher at a steep institutional subscription price. However, this worthy initiative had to stop after only 5 years of existence.

Where do we stand now? Some lament the passing of scientific journals after 350 years of existence. Are they being unduly pessimistic? No week and even no day goes by without my being pestered by spam e-mail by editors and publishers of predatory journals—even though I have been retired from active science for the last

20 years. Are Open Access journals a cure-all? Are platforms for preprints a viable solution? I am no prophet. I can only state that some action is needed to rescue scholarly publication as it has existed for centuries: it is as necessary as the air we breathe.

Tradition is beautiful, innovation is lovely. But how to strike a balance? My career moved pendulum-like between the two. In terms of changes I have initiated, the first that come to mind concern education rather than science proper. When I arrived at the University of Liège in 1970, the Faculty of Sciences was steeped in tradition. Its voting procedure followed seniority: the eldest professors who, in addition, had already discussed it on the phone, voted first and thus stated their position; the others could but follow suit and cast an identical vote. I introduced instead voting alphabetically, with the Dean picking at random the initial letter for the name of a faculty member called upon to vote first. This seemingly minor change was a small revolution. There were others, also procedural.

At that time of 1970, exams in Liège at the end of the academic year, for at least a month, were administered orally by professors in person to each of their numerous students. This burden was traditional and everyone heeded it. I had little merit in starting to overturn it: I came from the US, where I had worked with the Educational Testing Service in Princeton and I introduced regular, more or less monthly testing with multiple-choice questions and computerized grading. It spread slowly through the university.

Is this why when I returned to my own country, France, in 1986 I was viewed as a dangerous reformer, as a foreign (American) agent and thus hindered or resisted at every step? I was countered by colleagues intent upon upholding their privileges, especially the wielding of power through the indeed traditional tools of cronyism and patronage.

What indeed is beautiful about tradition and lovely about innovation? To follow tradition is a pleasure. One is aware of maintaining it, which is comforting and reassuring. To emulate one's distinguished predecessors, one's mentors especially, and to publish similar papers in the same journals is exhilarating: I vividly recall the enthusiasm of Edgar Lederer (1908–1988) after he saw a communication of mine in *JACS*.

To innovate is a joy, of finding oneself in a no man's land, of treading on new ground and of starting to explore the new territory. I had this feeling a few times— too few, I ought to have been more daring.

There was the time when everyone in nmr was rushing to do carbon-13, as soon as commercial spectrometers acquired that capability, in the early Seventies. I did likewise for a short while before I had the guts to innovate instead: my coworkers and I pioneered sodium-23 nmr, where we were rewarded at every step.

To turn from instrumental innovation to problem-oriented innovation, we realized that the Na^+ sodium cation has about the same radius as the Ca^{2+} cation. Hence, it can be used to explore calcium-binding sites in proteins—which we did.

The verb "to innovate" was rarely used in both French and English until the sixteenth century. It was very much an acquisition of the Renaissance, together with renewed curiosity in the working of the natural world. In his *Essays* (1597), Francis

Bacon (1561–1626) wrote explicitly of innovations. This deep thinker of genius wrote that we ought to "we make a stand upon the ancient way, and then look about us, and discover what is the straight and right way, and so to walk in it."

I walked in it joyfully when we started using clays as assists in chemical transformations: catalysis using clays improved many of the key reactions in organic chemistry, the so-called name reactions. To do that work to me was intensely satisfying—definitely an understatement.

My time as a scientist, 1960–1999, occurred while science was dominated by the United States. From the end of World War II until the end of the century, American hegemony was in full force. During that period, about 50 Nobel prizes in chemistry went to Americans or foreigners working in American laboratories. But what were some of the factors for such overwhelming dominance?

Of course, one of the reasons was that English served as the language of science, the lingua franca worldwide. Which at the time of writing is still the case. Will Mandarin dethrone it and, in the affirmative, when will that be? We chemists know that Asians brought up on ideograms have an inborn advantage in the idiom of chemistry based on formulas, i.e., iconic small diagrams.

What were, in my opinion, some of the other reasons for American dominance of science and chemical science? I would put in second place the "can do" mentality. Americans are doers. They do not give up easily, do not get discouraged and are willing to try again and again.

There is also the "hands-on" attitude. In contrast to Europeans, the French say, who make a sharp distinction between manual work and intellectual work, the former being judged menial and secondary, Americans do not shy from dirtying their hands and, for instance, building apparatus rather than relying only on purchase of commercial instruments.

Competition is another important factor. Brought up on team sports, Americans as a rule are highly competitive, they want to be the best, individually and collectively. Which often goes with being assertive.

As a cognate of competition, I need to mention discussion and controversy, essential ingredients to the advancement of knowledge. Americans basically have no respect for authority of any sort and delight in challenging it.

Yet another factor, that goes with the "hands-on" approach, is ease at handling money. When I first came to the US, in 1962, I was impressed by the ability of Americans at budgetary planning. For instance, where in my country new scientific instruments came only with new buildings for science, i.e., much too infrequently, in the US it was much easier to obtain the necessary monies; plus, yearly running costs which, most unfortunately, in France were not provided for. A contrast, one might say, between a society of abundance and a society of scarcity.

Now that one can foresee China replacing the US as the number one scientific power, no doubt Chinese cultural elements will replace partly or totally the above short list.

I retired in 1999. Since then, I have kept an eye on science and new developments. What are some I did not anticipate and thus were surprises? What I think of them is another question.

The first change that I had not foreseen was multiplication of journals. With the advent of the Internet I knew that scientific communication would move online, journals included. My awareness, though, did not extend to such a great increase in the sheer numbers of journals: I underestimated greed from some commercial publishers and their continued bilking of university libraries.

Since I mentioned the Internet, like many other scientists I am nostalgic for the era when we had this magic tool for ourselves before it became widespread in the public at large and gave rise to social networks, a reality of our time, a delight and a scourge both.

I dwelt in a previous paragraph on the highly beneficial Americanization of science during the second half of the twentieth century. It had a couple of consequences that proved to be at least as deleterious as desirable: scientometrics on one hand, ranking of universities on the other.

The former brought us impact factors for journals, which allowed scientists to aim their papers for maximum visibility—definitely a good thing. But it has to be balanced against another reality, the opacity and radical newness of major new finds. Think of Gregor Mendel's (1822–1884) discovery of genetics and of its near-invisible publication in 1866. As another instance, I can cite Alexandru T. Balaban's (b 1931) application of graph theory to chemistry, published during the Sixties in the definitely low-impact journal *Revue Roumaine de Chimie*.

The radically new, the novel and truly innovative breakthrough, at first remains opaque and misjudged within existing fields of science. In addition, a scientist may refrain deliberately from publishing but find nevertheless a way to assert priority: an example is the *pli cacheté* device at the French Academy of Sciences which was resorted to by eminent scientists such as Louis Pasteur (1822–1895) or Henri Poincaré (1854–1912).

I turn now to the ranking of universities that the so-called Shanghai Ranking put in the forefront: it is important to university administrators and other organization people more than it is to us scientists. I see it predominantly as the highly public endorsement by the Chinese of American rankings, well-known during the second half of the last century, that changed very little then or since: to mention its fairly representative 2017 appearance, Harvard—Stanford—Cambridge—MIT—Berkeley—Princeton—Oxford—Columbia—Caltech—Chicago—Yale—UCLA—U of Washington—Cornell—UC San Diego. As an aside, I am proud of having taught, from being on the faculty or as a visiting professor, in no fewer than four of those leading institutions.

But let me turn back to changes I failed to foresee. The new system of lab-work, featuring large research groups and multiple (>4) authorship, has replaced the traditional format of dual authorship by the mentor and graduate student (or post-doc). This reflects a new organization of labor, the change of chemistry from a craft and apprenticeship from one of its masters to a small industry. Is it for the better of for the worse, is not for me to say. But it is a major change, with the concomitant increase in production, i.e., in number of publications—which helps to explain journals proliferation. I see it, to restate a point already made, as the result of a markedly increased influence of industrial chemistry: technological applications nowadays rule over the discipline.

To sum it up, I failed to foresee the abruptness of the changes. They thus amount to mutations rather than mere evolutions.

Me-Too Chemistry

19

Are chemists gregarious? This might be the thought of an outside observer, at one of the annual meetings of the American Chemical Society. These meetings draw routinely 18–20,000 participants (Fig. 19.1).

And could there be probable cause for such gregariousness? Actual or perceived loneliness in laboratory work? The need for a feeling of brotherhood, that may stem from shared group values, such as the technical language—graphic formulas for instance—that sets them apart from other scientists, with only mathematics as a common language? The sulfurous alchemical heritage that to some extent made them social outcasts?

These questions raise themselves from collective behaviors that may seem a little weird. My first example, not far remote in time (the Eighties), is synthesis of quadrone. This natural product, from the class of sesquiterpenes, shows antitumoral activity. Hence, its synthesis became desirable, so that it could be turned into a marketable drug by pharmaceutical companies.

One might guess that, accordingly, two or three academic laboratories met with such a goal. The answer, rather than just two or three, is two or three *dozen*. Scientists resemble other people in sometimes choosing imitation over creation. They can fall victims to fashionable trends, hence espouse me-too chemistry.

I detest the term technoscience and what it stands for. It may have been coined by Bruno Latour (b 1947). In any case, it was used by his followers, the self-named postmoderns, in reference to the science-technology alliance bred by consumer society. In their minds, the border between science and technology had become so porous as to have become inexistent. To them, science rather than a purely intellectual activity, had become tainted with the utilitarian and commercial interests of technology. Thus, their knee-jerk reference to science as technoscience amounted to a term of abuse.

Very often, me-too chemistry—fashionable trends in chemistry—stems directly from technoscience. During my career, I witnessed—from outside, I hasten to say— several such episodes; during which chemists flocked *en masse* to try and pocket profits, whether in academic or social position, or just a windfall of plain material

Fig. 19.1   Drawing by Escher
(fair use)

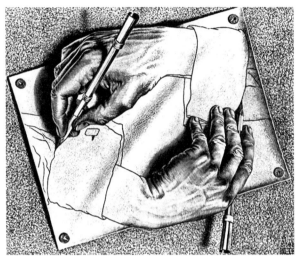

benefits. Such episodes had a thin positive lining in bringing together chemists from academia and from industry.

There was polywater, a craze from the Sixties (well-documented by Felix Franks, (1926–2016)) originating in the Cold War and the US-USSR military rivalry. It occurred during my time at Princeton, half-a-dozen of my colleagues in the Department of Chemistry succumbed to it.

There was, a few years later, in 1989, cold fusion. It originated in another department of chemistry at the University of Utah. And resulted in its two progenitors, Martin Fleischmann (1927–2012) and Stanley Pons (b 1943), dropping out of science altogether.

My country, France, was not immune to such spikes of pathological science— after all it had been, earlier in the twentieth century, the site of Blondlot N-rays. It became, for a short while (in 1988), the locale of Jacques Benveniste's (1935–2004) *mémoire de l'eau*, an episode of pathological science, nurtured by the French affection for homeopathy; and by its later revival and support from Luc Montagnier (b 1932), who was awarded the Nobel prize in medicine in 2008.

I well recall my indignant discovery of *Le Monde* not only having taken seriously Benveniste's allegations but publishing them on its front page as a major scientific discovery. Fortunately, my friend Jean-Marie Lehn (b 1939) dashed off a violent rebuttal, published the following day. At that time, I was a professor at the *École polytechnique*. Given the prestige in France of that institution, Benveniste, through my brother-in-law, solicited my support; which of course I refrained from offering.

Each of these cases was a mass phenomenon of chemists rushing to find possible rewards from the novel phenomenon, after of course making sure of its authenticity. In this chapter, I shall recall my impressions of synthetic organic chemistry as it too amounted to a fashionable trend, sucking quite a few research groups into the maelstrom. I shall also delve into a field of definitely technoscience, shape-selective catalysis within zeolites.

Why on earth would leaders of research group in synthetic organic chemistry have so little imagination as to set themselves this single goal of quadrone as their target molecule?

Total synthesis of natural products, in some of its group leaders, shows a need for self-glorification. Its rhetorical devices are easily recognized, being ritualistic. There is the homage to R. B. Woodward, who gave distinction and prestige to the field, crushing it in the process under the weight of his genius. Borrowing a propaganda trick from the media, synthetic organic chemists are prone to liken their achievements to those of mountain climbers scaling formidable peaks. And indeed, just like Mount Everest, they claim distinction from following in Woodward's footsteps. His followers are often just imitators. They attached themselves to total syntheses of molecules, chosen for the combination of a complex architecture and a marketable biological activity. One can feel the despair from a published lecture invoking the names of recent Nobel prizewinners in chemistry, with the transparent hope of earning a similar accolade. The intellectual weakness of such name-droppers is palpable. A good rule of thumb, whenever a giant is invoked, is to ascertain the weakness its is meant to cover.

Woodward, Mount Everest, Nobel prizewinners: what else? There is as a rule the homage to Nature with a capital N. It is likewise routine. It is often complemented by mentions of the biological activity and of the use, actual or potential, as a drug. The synthetic organic chemist thus projects himself—it is a He as a rule—as a benefactor to mankind; when actually he has trained the graduate students and postdocs in his group for future work in industrial laboratories, within the fold of pharmaceutical companies to which he provides skilled labor. Just like chemical science as a whole, pharmaceutical companies emanate from the industrialized world, led by the US, China, Japan, the UK and Germany, Switzerland.

I'll mention two examples only. They show the amazing degrees of conformity, of dominance by the market following Adam Smith, of the lack of intellectual creativity—as a rule these total syntheses are recipes liking together a series of name reactions, each with its anticipated purpose.

My first example is taxol.

My second example is quadrone.

I watched from close quarters the unfolding of the taxol story, during the Seventies and Eighties especially. A person very close to me had her cancer treated by this drug, in the form of *taxotère*. I was a close friend with two of the leading investigators who brought the drug to patients in, not only a practical way, one also respectful of a threatened natural species: I was a close friend of Pierre Potier (1934–2006) since the early 60s; and I befriended Bruce Ganem (b 1948) while visiting at Cornell during the 80s.

Pierre Potier (1934–2006), first: when I prepared my doctorate with Professor Edgar Lederer at the CNRS Institute for Chemistry of Natural Products, in Gif-sur-Yvette, Potier seconded Maurice-Marie Janot (1903–1978) in managing the half of that Institute with a staff of pharmacists. Trained himself in pharmacy, PP had wed the owner of a pharmacy in the Paris area. Tragically, Marie-France his wife died still very young, in 1968, at the Cancer Institute in Villejuif, from breast cancer. Pierre then made it his lifelong goal to defeat cancer. Indeed, we owe him two outstanding anti-tumorals, *taxotère* (Rhône-Poulenc) and *navelbine* (Pierre Fabre). Pierre was unconventional, spoke his mind and was a leader of French organic chemistry during his—too brief—life.

My recollections of him include visiting his vast luxury apartement in the seventh arrondissement of Paris, after he had remarried Odette; several joint stays in south-western France, Pierre was very fond of the Gers *département* and of its Madiran wines. This geographic polarization led to a friendship wih the political leader François Bayrou (b 1951) which in turn got Pierre appointed (1994–1996) director of French higher education and research, in the Ministry of Education. As yet another detail, PP collected clocks of the *cartel* kind.

Now to some of the highlights of the taxol story—it belongs with this chapter on me-too chemistry since, during the period 1960–2000, journals of the American Chemical Society alone published *thousands* of papers on taxol. K. C. Nicolaou (b 1946) and Samuel J. Danishefsky (b 1936) alone published about 50 papers on aspects of the synthesis of taxol. In 1999 alone, the drug Paclitaxel marketed by Bristol Meyer Squibb earned $1.5 billion in sales, it is not irrelevant to mention.

Earlier on, the first synthesis of taxol in the chemistry department at Florida State University, in Professor Robert A. Holton's (b 1944) group, ran into a conflict, between the public defense of a Ph. D. dissertation and the confidentiality demanded by the application for a patent.

Taxol was first isolated from the bark of the Pacific yew (*Taxus brevifolia*) whose survival as a species became threatened by its therapeutic use: initially, treatment of a single patient for ovarian or breast cancer demanded sacrificing an entire tree. A bird, whose habitat was also the Pacific yew, the northern spotted owl, also became threatened.

Both my friends, Pierre Potier and Bruce Ganem, developed ways to obtain taxol without having to kill entire trees. Pierre found a hemi-synthesis of taxol from needles of the abundant English yew. Bruce did likewise from needles of nursery-grown ornamental yew trees that take only 3–4 years to regenerate; in collaboration with Natural Products Inc. (NPI). Drying of the biomass within 5 h of harvesting produces up to 100 t of plant material per day, about 10% in paclitaxel. It is then shipped to Mexico for extraction there. Further isolation and purification is performed in Vancouver, Canada.

Half-hearted or full-hearted? Preparing taxol from an ingredient of yew needles characterizes both Pierre's and Bruce's accomplishments. They qualify as hemi-syntheses, rather than total syntheses.

What is the difference and does it matter? Purists, i.e., apostles of total synthesis, invoke the tradition. Going back to the nineteenth century and to the French chemist Marcelin Berthelot 1827–1907), total synthesis translates into synthesis from scratch, i.e., from the elements. Such a dictum or edict is of course at a distance from the actual practice of synthesizers: they will use as starting material a molecule that, at some time, may indeed have been made from the elements.

I submit that the 'total' epithet is meant as 'made in academia,' 'art for art's sake,' i.e., supposedly as a gratuitous pursuit, at a distance from any lucrative concerns. Which amounts to an oxymoron, since these laboratories and their research groups thrive, conversely, on support from the pharmaceutical industry—in terms of equipment, consultancy fees for the group leader and jobs for the young scientists thus trained. The expression 'total synthesis' is a smokescreen for these crucial aspects.

Ethically-speaking and not only in terms of environmental conservancy, the hemi-syntheses of taxol by both Potier and Ganem are considerably more honest, both morally and intellectually.

Now to quadrone: it dates back to the Eighties: there was a time when in the US, about 60 research groups in chemistry worked on the total synthesis of quadrone. What is quadrone? Why was this particular molecule deemed important? What kind of an impact has it had in pharmacology? What have been the lasting consequences of such a massive investment in personnel and resources? How does it reflect on synthetic organic chemistry?

Quadrone is a sesquiterpene molecule produced by a fungus,. Scientists from the W. R. Grace company reported in 1978 its antineoplastic potential. This is what triggered a massive effort, on the part of academic laboratories to achieve its total synthesis. The obvious hope was that it would be turned into a drug, in similar manner to taxol, which you have just read about; and that pharmaceutical companies would benefit from the forthcoming sales.

All of that came to nothing. The reason was that quadrone was found unsuitable during the Eighties as a potential anticancer drug. Its story indeed reflects poorly on synthetic organic chemistry: that it needs for its output to be useful—financially useful—to pharmaceutical companies; put another way, that facilities in academic settings, i.e., nonprofit organizations are exploited for profit-making by commercial outfits. To put it another way, quadrone in its various total syntheses did not hold enough intellectual challenge or satisfaction to justify it by itself. Which puts synthetic organic chemistry in a bleak light.

Now, almost a half-century later, where does synthetic organic chemistry stand? It is in the process of being relegated to the dustbins of scientific history: artificial intelligence (AI) is taking over, following upon the lead of E. J. Corey's (b 1928) retrosynthetic analysis. Computers are replacing human minds whether in the overall planning of a synthesis or in the choice of the various steps, i.e., the name reactions resorted to for each of those. In this manner, me-too chemistry has rushed into a dead-end.

How did synthetic organic chemistry fare, in the eyes of other chemists? For sure, the exalted self-opinion of its status and merits did not meet with overwhelming approval. The niche it had closeted itself in to some extent was also a ghetto. Of course, its self-ranking at the top of the totem pole rankled other groups among chemists: foremost, those it had expelled from the number one position, the physical chemists and the physical organic chemists. Synthetic organic chemists were viewed, during the last few decades of the past century, with some commiseration, as being too highly specialized for their own good—with pharmaceutical industrial laboratories as their major and single job outlet.

As late as 2000, a leader in total organic synthesis published a comprehensive review, amounting to self-promotion addressed to the Nobel Committee. This document states that art and science continues to fuel the drug discovery and development process with myriad processes and compounds for new biomedical breakthroughs and applications.

There were chemists though, close enough to synthetic organic chemistry to have gained familiarity and expertise. They viewed it with a jaundiced eye. They saw that its heyday had passed and that it was time to move on. As we saw, the Eighties saw a flurry of total syntheses of both quadone and taxol—the examples I chose. As early as 1990, Dieter Seebach, a professor at the Swiss Polytechnic in Zürich, in a comprehensive review of the field, could write: "almost everything the study of unusual molecules could teach us about the nature of the chemical bond has probably already been learned. In other words, all the most important traditional reasons for undertaking a synthesis-proof of structure, the search for new reactions or new structural effects, and the intellectual challenge and pride associated with demonstrating that "it can be done"-have lost their validity." He added, "the exciting synthetic targets today are no longer molecules to be prepared "for their own sake"; instead, they are systems associated with particular functions or properties. Organic chemists are busy designing new materials." And "the effort will not have been wasted if I have convincingly swept away the one-sided but prevalent notion that organic chemistry and organic synthesis are mature sciences."

The lucidity he displays in that review article, is as appealing as well-documented: "The prophetic observations of Woodward in his famous 1956 essay on the subject of "Synthesis" remain just as valid today. On the other hand. it was left to a magician like Stork to propose a time scale: "So it is not surprising that organic synthesis is far from the level that many people assume. Progress is continuing, but there will not be any dramatic development. It is more like a glacier that gradually moves forward until it has finally covered an entire region, but it will still be centuries before synthesis has acquired the status that many people already ascribe to it today." D. Seebach, "Organic Synthesis—Where Now?," *Angewandte Chemie International Edition in English,* 29, 1320–1367, 1990.

By 2000, pharmaceutical laboratories indeed had turned away from the total synthesis of natural products of complex molecular architectures. A more realistic picture had been sketched by Professor Dieter Seebach (b 1937), of the Swiss Polytechnic in Zürich, as early as 1990—the Eighties were the heyday of total organic syntheses as me-too chemistry, that saw in particular between 1980 and 1990 no fewer than 16 elaborations of quadrone, a clinically-useless molecule as it turned out already at that time.

There were other downgrading attitudes, though, however necessary. I wish to bring into consideration a most worthwhile attempt, at rethinking the missions and purposes of synthesis, in the absolute, as it were. For this purpose, I can do no better than refer readers to the proposals by George Whitesides (b 1939), along these lines. He expressed them in 2018. The date is significant, Whitesides had ample time to reflect on the axiology of the synthetic endeavor, envisaged as a whole, from the outside, but by a well-informed and kindly-meaning outsider. A key question

Whitesides asks is: "is the training that students currently receive in organic synthesis the one that best prepares them for their future (and which may possibly be entirely different from their research director's past)?" Whitesides's thoughtful essay, however, is not only a requiem for organic synthesis, as a branch of chemistry having outlived its necessity and usefulness; as he writes, "a style that emphasizes technical proficiency and complexity rather than simplicity, breadth and utility."

Whitesides's argument is predicated by the key observation of the pharmaceutical industry and the American National Institutes of Health having moved away from cytotoxic chemicals used as drugs against cancers, to biopharmaceuticals, proteins primarily. Indeed, we are witnessing the end of an era. What is the next going to be? Whitesides answers that very question: "biological processes have become a major part of synthesis, from research demonstrations to commodities (. . .) these methods have been largely ignored by the community of specialists." Whitesides, three years ago, foresaw the predominant role of computers and information science in the future of organic synthesis, which we start to witness.

But why such a collective rush into a trendy area, as exemplified by organic synthesis? At least part of the explanation, I submit, comes from dominance of science by Americans during the second half of the twentieth century. After World War II, research universities came into being and led the worldwide research effort in the ensuing decades. During the Cold War, institutions responsible for funding research and awarding grants, such as the National Science Foundation and the National Institutes of Health, followed in the footsteps of major wartime projects such as the Manhattan Project or the manufacture of penicillin. They adopted military-type organizations: safety in numbers; the efficiency of a task force in gaining success. There was also distrust of individuals going forward by themselves. Not to mention the influence from sports such as American football: collective action vs individual dashes.

David Riesman (1909–2002) in *The Lonely Crowd* (1950) described how other-directed people, i.e., a majority of Americans, were responsible for the smooth functioning of modern organizations. The book also noted that a society dominated by the other-directed faces profound deficiencies in leadership, individual self-knowledge, and human potential. Which of course is consonant with science being run in those times by mass advances rather than progressing from individual forays.

I became aware of my own attitude, the latter, in the early Seventies after I had moved to the University of Liège. The physical technique I was an expert in, nuclear magnetic resonance (nmr), underwent a small revolution at that time: the advent of Fourier-transform data acquisition opened up carbon-13 spectroscopy to organic chemists. They rushed to take advantage of it, being so well suited to the study of organic molecules. I was very much tempted to do likewise, in order to keep up with the pack. But my instinct told me to look elsewhere—and we (my group and I) started instead sodium-23 nmr, of which we were pioneers and that brought us numerous rewards. Take-home lesson: do not follow others, go by yourself into the wilderness and you might be able to devise a new research area for yourself.

Zeolites provide another outstanding example of me-too chemistry, one I was not a participant in, a privileged observer only. Looking-up this term or equivalenty 'shape-selective catalysis' confirms its status: this sub-field is one chemists have rushed in for now more than half-a-century. Google Scholar shows 4010 hits in the Sixties, 11,700 during the Seventies, 18,500 in the Eighties, 48,100 in the Nineties and 102,000 during the first decade of the current century. Such a geometric progression, with a doubling approximately every 10 years, is reminiscent of Moore's Law for computers. It reflects a similar manufacturing and industrial background, which explains the involvement of academic chemists, moved by the pull of their industrial counterparts.

What are zeolites? How did I become acquainted with these minerals? Who were some of the protagonists and some of the leaders in this area at the border line of chemical science and its industrial and commercial applications?

Better if I rank my recollections chronologically. The first time I heard of zeolites was in 1960, when the World Congress on Catalysis was held in Paris. I was still a student then. The word "catalysis" appealed to me and I applied to attend this conference. My application was referred to Boris Imelik, who was in charge of the technical aspects and I was allowed attendance, with all charges waived. There were half-a-dozen of us young men—others were Michel Che (1941–2019), Jacques Fraissard (b 1934), Jacques Védrine (b), who all went on to make a name for themselves in Catalysis. The meeting was held at *Maison de la Chimie*, in its main auditorium. What was the function, of us young students witnessing the historic exchanges by leaders in the field of catalysis? We were in charge, during the discussion periods, with rushing and obtaining on the spot from each discussant to write down on a paper slip the text of his question or comment, for future printing in the Proceedings. To me, what made this conference memorable were the heated arguments between Russian and American delegates, led on one side by Nikolay Semenov (1896–1986) and on the other by the extraordinary and most articulate John Turkevich (1907–1988), whom I would meet again as a colleague at Princeton, at the end of the Sixties. One of the highlights of the meeting was the report of the catalytic activity of zeolites.

But first a word about the arguments between the Russian scientists and the Americans, led by John Turkevich, in the context of the ongoing Cold War. John would routinely win them. Clearly, he was more experienced in scientific discussions and more apt at stating relevant evidence—a most convincing debater. I came to appreciate his geniality, among other qualities. I learned later that, in addition to being an outstanding chemist, he was the son of the Primate of the Russian Orthodox Church in the United States. In addition, he was widely rumored at Princeton to be recruiting actively for the CIA. In 1960, he became the first science *attaché* named to the United States Embassy in Moscow, and in 1965 was the head of a delegation to the Soviet Union to establish scientific exchanges with the United States. The behavior I witnessed in 1960 heralded such a leading political role.

Coming back to zeolites, what then are they? The short answer is aluminosilicate minerals, both natural and artificial. The name zeolite, meaning "foaming stone," was given by the Swedish mineralogist Axel Fredrik Cronsted (1722–1765), in 1756. He observed that rapidly heating the material produced large amounts of steam from water adsorbed in it.

I came upon these minerals from their widespread use in the Sixties under the appellation molecular sieves. They present pores on their surface and internal chambers. The former connect by channels with the latter. They were used, in particular, for keeping dry organic solvents. With a certain type, the pores had dimensions such as to welcome water molecules. Use of molecular sieves could not have been easier, one poured a small amount into a bottle of the organic solvent. This was easier and less dangerous than the former procedure, consisting in extruding filaments of metallic sodium into the solvent—a fire hazard. And the molecular sieves could be regenerated very simply, by drying in an oven prior to renewed use.

Their main application, exciting to many a chemist and that became a chapter in itself within chemistry—and me-too chemistry—was in so-called shape-selective catalysis; of which I had been given a foretaste at the Paris conference in 1960. Presentation of the future process at that conference was very quickly followed by a most impressive application to petrochemicals: This very rapid first marketing of a zeolite-catalyzed process in 1962 occurred in gas oil cracking, the largest volume catalytic process with a worldwide capacity of over 40 million barrels per day. It made use of large pore zeolites with pore openings described by a 12-ring of tetrahedra.

Research on zeolites blossomed very quickly. As of December 2018, 245 unique zeolite frameworks have been identified, and over 40 naturally occurring zeolite frameworks are known. Every new zeolite structure was submitted to the International Zeolite Association Structure Commission and received a three letter designation. By the mid-Eighties already, "the advent of high silica, medium pore sized zeolites had opened a myriad of new opportunities in synthesis, characterization and catalysis." In particular, a synthetic (artificial) zeolite known as ZSM-5 stole the limelight, with its 10-membered pore opening. The acidity of ZSM-5 served for acid-catalyzed reactions such as hydrocarbon isomerization and the alkylation of hydrocarbons. One such reaction is the isomerization of meta-xylene to para-xylene. This "wonder catalyst" discovered and made by Mobil converts methanol to gasoline.

More generally, shape-selective catalysis is the molecular-sieving occurring during a catalytic reaction: it discriminates between the reactant, the product or the transition state species in terms of the relative sizes of the molecules and the inner chamber space where the reaction occurs.

In the context of zeolites, I should mention one of the leaders in that area of chemistry, whom I got to know well and to befriend. This was Eric Derouane (1944–2008). When I arrived at the University of Liège in 1970, he was there, a rising star; and he took me under his attentive care. The mutual sympathy came from his having worked in John Turkevich's lab at Princeton, receiving an M.A. in 1966. The reason, as it turned out, was his belief that I might be influential in furthering his future career in Liège. Accordingly, not only did he take me under his wing, introducing me to the specificities of life in Belgium, he and his then wife, Claudine, hosted me for weeks in their apartment within walking distance of the Sart-Tilman, the then new campus.

Rising star? Derouane was being helped along by his mentor, Louis D'Or (1904–1989), who had recognized his excellence and meant for the University of Liège, in which he was highly influential, to benefit from such a talent. However, Professor D'Or suffered from a poor health and took early retirement, shortly after I had arrived in Liège. He would be replaced, not by one, not by two, but by half-a-dozen of his former associates who had yet to become appointed full professors: Jacques Collin, Jean-Claude Lorquet, J. C. P. Mignolet, Jacques Momigny and Pierre Tarte. Taking advantage of their greater seniority, they effectively ousted Eric from the University of Liège. He went to the (Catholic) young university in Namur, where he started building an empire. Subsequently, he returned to Princeton in 1982–1984, but not to the University, as head of the Exploratory Catalysis Synthesis Group at Mobil Research & Development Corporation, Central Research Laboratory, Princeton, USA—the very laboratory where the famous ZSM-5 zeolite had been born. In 1995, Eric was made a professor at the University of Liverpool, while retaining his appointment and responsibilities in Namur. From the early Seventies, Eric worked on the ZSM-5/MFI new zeolite, leading to a 30 year collaboration with the above mentioned Jacques C. Védrine. In 2003, Eric was appointed to a Gulbenkian Foundation-funded Professorship at the University of Algarve in Faro, Portugal where he was Director of the Chemical Research Centre. My own research on chemistry using clays—cousins of zeolites in being also aluminosilicate minerals—made me aware of other eminent scientists having contributed to the zeolite chapter of chemistry, such as Colin Fyfe (b 1942) and Sir John Meurig Thomas (1932–2020).

But I do not want to let go of the topic of zeolites without a mention of the 1994 laundry war between Unilever and Procter and Gamble (P&G). The former giant corporation started it, rather inadvertently, being too quick to introduce a bio-inorganic enzyme-like component, devised by scientists at the University of Amsterdam. The blunder came from their marketing staff, who pushed for its introduction, without waiting for a full battery of tests to be conducted. At that time, P&G owned 36% of the world market in detergents versus 23% for Unilever, which explains this attempt at catching up with a brand-new product. After a few washes with Persil Power, launched in May 1994, clothes first started to lose color and then tore under any stress. I was hired as a consultant by P&G and spent a week, with about two dozen outside scientists at the P&G European research laboratories near Brussels. P&G won easily that war, Unilever was forced to withdraw their brand-new product, the Persil Power. By that occasion, I found out how complex the formula of a laundry detergent is! At that time already, it included a zeolite (a builder, in the jargon of the trade) to soften water to avoid the environmentally damaging effect of phosphates, viz. eutrophication of streams from the cumulative outflow of washing machines.

A word in closing about the expression 'me-too chemistry'. It has antedated by a few decades the current feminist revolt against sexual harassment by men in the workplace. The 'me-too' adjectival expression seems to have originated towards the end of the nineteenth century. I first heard the phrase 'me-too chemistry' during the Sixties. It was uttered by Hugh Felkin (1922–2001), who worked in the same

Institute of Natural Products in Gif-sur-Yvette, where I had worked myself on my doctorate. A first-rate chemist, an expert in mechanistic organic chemistry, he was a lucid—often deprecating—observer of the contemporary scene.

What Have I Achieved?

20

In this chapter, I revisit some of my contributions to chemistry for their aftermath and influence. In addition to my development as a scientist, also as a person they brought me strong friendships with a remarkable cast of characters.

I start with the epochal and spectacular reaction that André Cornélis (1941–2021) studied in his doctoral work at the University of Liège. We published it in 1975 as a communication to *JACS*. I described in a previous chapter its presentation at a GECO conference in Locguénolé, Brittany and the enthused reception from fellow-chemists Breslow, Mislow and Seebach. We had named it a *cascade* reaction on account of the successive intermediates it went through before its final, resting state.

Some years later, Horst Prinzbach (1931–2012), a professor at the University of Freiburg in the Black Forest, wrote with the belief that the name was original to our paper, requesting permission to apply it to a transformation achieved in his research group. Of course, I acquiesced. Their work appeared in 1982 in *Berichte*.

But the great fortune of that name—unacknowledging its origin in Liège—came from synthetic organic chemists. I shall single out Professor Kiriacos C. Nicolaou (b 1946) for exploiting the efficient elegance of cascade reactions. He credited their rediscovery to Lutz F. Tietze (b 1942) in the 1990s. Of course, the original cascade reaction, without such a name, is the epochal and spectacular total synthesis of tropinone by Robert Robinson (1886–1975) in 1917.

In 2003 a group headed by Michael H. Howard—from the Central Research Department at DuPont—termed the Liège reaction "fascinating". They praised our elucidation of the detailed course of the transformation. In 2006, Dieter Enders (1946–2019) wrote that "catalytic cascade reactions can be described as biomimetic, reminiscent of reactions that occur in biosyntheses of complex natural products." Nature makes extensive use of cascade reactions, wrote a group of Chinese chemists in an *Accounts of Chemical Research* 2012 paper relating their reliance on such transformations for building elaborate architectures. Another Chinese group published in the same journal in 2015 their use of catalytic asymmetric cascade reactions.

P. Laszlo, *A Life and Career in Chemistry*,
https://doi.org/10.1007/978-3-030-82393-1_20

This was all gratifying. Did we do better than just coining a name? At that time of the Seventies, no doubt rooted in my nmr work, my predilection went to chemical physics. Michel Jauquet and I studied molecular dynamics and correlation functions. As Lucretius tells Memmius, *Nunc quae mobilitas sit reddita materiai corporibus, paucis licet hinc cognoscere* (let us now learn in a few words how fast atoms move). All that came to an abrupt halt when Michel was killed by a hit-and-run driver. It broke my heart.

I turn now to the immediate aftermath and medium-term posterity of our work with Armel Stockis (b 1951) on norbornenone reacting with iron carbonyls— remarkable for an amazing stereoselectivity, highly unusual outside of biology. It led foremost to strong and often lifelong friendships with a number of persons, Armel himself taking pride of place.

Its closest follow-up was with Armel serving a post-doctoral stay in Roald Hoffmann's (b 1937) laboratory at Cornell; successful enough for Roald to mention it in his Nobel lecture. Armel's assignment in Ithaca was to explain our Liège results using molecular orbital (MO) theory. If I may be technical for just half-a-sentence, the impressive stereoselectivity of the transformation arises from the ethylenic LUMO (lowest unoccupied MO) entering the reaction with its largest lobe β to the iron in the product metallacycle. Which Armel and Roald's *JACS* paper in 1980 points out. As Virgil had written (*Georgics*, II) *Felix qui potuit rerum cognoscere causas* (Happy is he who has been able to understand the causes of things.) Already in 1976, another future Nobelist, Noyori Ryōji (b 1938, my near-exact contemporary), had cited our 1974 paper in a review chapter that dealt, among other topics, with cyclopentanone formation from iron carbonyls and strained olefins.

Our work on iron-carbonyl-induced coupling of ethylenics into cyclopentanones was closely paralleled by Edward S. (Ted) Weissberger (b 1941) at Wesleyan University, in Connecticut. He offered Armel another postdoctoral stay in his lab. As a consequence, I also visited Middletown several times and the lovely Wesleyan campus there. Ted and I wrote jointly an article in *Accounts of Chemical Research* (1976). I hosted stays in my Liège lab for a couple of Ted's coworkers, most notably William L. (Bill) Holder (Wesleyan Class of 1975) who, afterwards became editor for decades of the *Wesleyan University Magazine*.

At Wesleyan I made friends with the remarkable Max Tishler (1906–1989), who chaired the department after a stellar industrial career at Merck Sharp and Dohme in Rahway, New Jersey. Max was not only a chemistry genius, he also had a very sound judgment about people. I was lucky for a couple of my visits to Middletown to coincide with the Peter A. Leermakers Symposium: Max had started it after the untimely death from a car accident in Mariposa, California of a highly promising young faculty member—much later in life I found myself sharing with my wife Valerie a house in Mariposa and spent most winters there. Max invited chemistry luminaries to lecture in the Leermakers, the likes of Geoffrey Wilkinson (1921–1996), Al Cotton (1930–2007), Albert Eschensmoser (b 1925), Bill Lipscomb (1919–2011), Derek Barton (1918–1998), … I was lucky to interact with them during some of my stays in Middletown.

Paul Haake (1932–2011), an organic chemistry colleague who went on to create there a Department of Molecular Biology and Biochemistry, told me of the Sierra Club and of the trips it led into remote and magnificent parts of the USA. Thus, Martine my then wife and I took part, during the summer of 1976, in a memorable trek through the wilderness of Wyoming—a long-lasting friendship ensued with Coleman Citret (1937–1984), an orthopedic surgeon and amateur flutist, who subsequently introduced us to his lovely family in the San Francisco area.

While at Wesleyan, I also met and became friends with Phyllis Rose (b 1942) who taught in the Department of English and was an essayist in the mold of Virginia Woolf, whom we both admired. She had a lovely house adjoining the campus, where she lived with her young son Ted. She recommended that I read *The Mirror and the Lamp* by M. H. (Mike) Abrams (1912–2015) whom I would later meet and befriend in Ithaca: he had shielded at Cornell Vladimir Nabokov from colleagues turned into enemies by Nabokov's venomous tongue and pen. She was delightful company and also introduced me to her colleague in English and close friend Annie Dillard (b 1945). She would marry another Frenchman, Laurent de Brunhoff (b 1925), a relative of my NSE classmate Thomas Ginsburger (1937–2020), the son of the Resistance heroes Pierre Villon (1901–1980) and Marie-Claude Vaillant-Couturier (1912–1996), and they would leave the Northeastern US to spend most of their time in Key West.

In closing this section I return to Armel Stockis: over the subsequent decades we remained in relatively close touch. He worked for many years in the pharmaceutical industry, lived in Brussels and is about to enjoy retirement in an incredibly handsome Alpine valley near Briançon.

My work with Ed Engler (b 1947), on the precise nature of the interaction of a polar solute with benzene solvent molecules, also led to remarkable and precious friendships. Not to a flurry of publications, though: rather than being the beginning of a long story, it was an endpoint and settled the matter. It was a self-teaching venture into the devising of a model. True, one can find later (1988–1992–2001–2007) references—about 150—to our 1971 paper; but its main importance to me was the discovery of Japan and its culture, to which I shall return at some length.

But first, a word about the further career of Ed Engler. After I left Princeton, he enrolled in Paul Schleyer's research group. He did beautifully in it, applying himself to Sandy Balaban's (b 1931) graph theory. He also got from Paul the urge to get to the top. He went to work as an IBM research scientist, in San Jose, California. He and his wife Peggy still live there. He was at the forefront of research on novel superconductors during the Eighties and the Nineties—for a while he owned the world record in superconductor performance; he then moved on to optical data storage. In the meanwhile, he rose in managerial ranks to become one of the directors of physical science in the IBM Almaden laboratory. William E. Moerner (b 1953), who also worked there, mentioned him by name in his 2014 Nobel lecture.

Now to my Japanese colleague and dear friend Nakagawa Naoya (b 1926). He contacted me when Ed Engler and I were scrutinizing the dynamics of camphor in benzene solution. Then at the University of Electro-communications in Tokyo,

Dr. Nakagawa worked out a model for that interaction. Thus, it was natural for me to let him know, in the early spring of 1979, that I would be visiting Japan at the end of April and would be going through Tokyo. I added that it would be a pleasure to meet with him.

What I did not expect was an invitation to stay in his home in Tokyo with him and his family, consisting of his wife Yukiko, an ophtalmologist, and their daughter Izumi. At that time, it was extremely rare to receive such an invitation into a Japanese home—quite a privilege. Indeed, when together with George C. Levy (b 1948) and his wife Linda our Aeroflot flight from Khabarovsk landed in the northern Hokkaido island, to our delighted surprise we had a welcoming party, consisting of Yukiko's sister and her son who lived in the area. They drove us to the railroad station in Sapporo and got us into a night train to Tokyo.

What are my main recollections of my first stay with and at the Nakagawas? Most memorable was their daughter Izumi's, then 14 or so, enacting for me the moving tea ceremony. But predominantly, Naoya's detailed initiation into the highly refined Japanese culture and mores. He and his wife lived in a suburb at a distance of an hour or so by suburban trains (two or three in succession) from the center of Tokyo. They took me to see the sights. Naoya accompanied me to and from the main campus of the University of Tokyo, where I was invited to lecture by Professor Ōki Michinori (1928–2016)—whom I had first met in Princeton.

This stay within the fold of a Japanese family blossomed into a lasting friendship. Some years later, Professor Nakagawa sent for a postdoctoral stay in my Liège lab (1997–1998) a coworker of his, Dr. Okada Shin-ichi. Dr. Okada made at least one memorable foray, in terms of prebiotic synthesis of biomolecules on clay surfaces. However, to my lasting regret, it occurred close to the end of his stay and he was unable to reproduce this key finding.

I returned to Japan in the mid-Eighties, with Valerie my wife, in April 1986 as a visiting professor at the Institute of Molecular Science in Okazaki, in the research group of Professor Iwamura Hiizu (b 1934) and his senior coworker, Dr. Sugawara Tadashi. The following August, my soon-to-be second wife Valerie and I returned to Okazaki, and we took some sightseeing trips. Another trip to japan in February 1989 took us to the Hiroshima region as guests of Mitsui Petrochemicals nearby. This company did not offer me a consultancy though. In any case it was a long shot.

My most recent visit to Japan started in Tokyo the first week of March 2015. Even though the country had visibly changed since the 1980s, it still offered a rich flavor. I stayed first in an hotel in the Meguro district. Afterwards I moved to the Nakagawa's, in Tsurukawa, a remote suburb.

To help me find my way there, the Nakagawas had sent me their grandson, Naoto, 18. I met not only Naoya and Yukiko, but also their daughter Izumi. She came to assist her parents with their foreign guest. I had not seen her since 1979.

Izumi in 2015 was about 50. In 1979, she had performed the tea ceremony in my honor; she was then learning it. In 2015, after I settled into the guest bedroom, she changed into traditional garb and again enacted the tea ceremony as a welcoming gift: this privileged moment erased the passage of time, no less than 36 years. Naoya was then 88; Yukiko a little younger. Both were in fairly good health but suffered from deafness—fortunately not too severe. However, Naoya showed his age, was less communicative, avoided the stairs in subway stations in favor of elevators—but still ate with a hearty appetite.

On Thursday March 12 2015 I met with Dr. Okada Shinichi. Passage of time: he had come 17 years earlier for a post-doctoral stint in my lab in Liège, two years before I retired. Two or three years later, he accompanied the Nakagawas, who were touring Europe and stayed for a few days in our house in Sénergues. We fell into each other's arms, he was obviously delighted to see me again; and happy that I had let him know I was coming to Tokyo. After lunch in a fancy restaurant, Yukiko, Naoya, Naoto and I repaired to his lab at Japex, a company into oil exploration. The Japex Research Institute is located in Makuhari, like Obeida and facing it in the north of Tokyo Bay: an entirely new district reclaimed from the sea, where the Tokyo Conference Center is located, relatively close to Narita International Airport. Dr. Okada showed us around the Research Center, then drove us to a nearby hotel, an American-style *palace*. We had earlier driven Naoto to the Makuhari station, where he started a more than two hours journey to return home. He had come, not only to lend a hand to his grandparents but also to help me with my bags. During the subway ride journey from Shinjuku to Makuhari, he had showed me, in a very kind and expert way, an origami folding that his grandmother had suggested—a little spinning top made of four sheets of paper, of different colors, embedded in one another.

Friday morning, Okada Shinichi came to pick us up with his car, a big and very comfortable sedan, a Mazda Premacy, for an excursion he had organized in the Honshu province, northeast of Tokyo. He wanted to show me the results of the disaster—almost exactly three years earlier—at the Fukushima nuclear power plant, from a devastating tsunami that had killed 20,000 people. We drove on freeways for 200 km until the immediate vicinity of the struck nuclear facility. During the last 20 or 30 km, illuminated panels on the roadside showed the ambient radioactivity (in micro-Sievert per hour): it went from 0.2 about 50 km away to a maximum of 5.4 about 15 km from the power plant. The road was flanked by fields, almost totally covered by long rows of large black garbage bags filled with surface layers of soil, scraped jointly with the fallout of radioactive dust. What would this material become? Be disposed in the ocean in a few decades?

We approached the crushed power plant, as close as it was allowed, a km or two. Then we drove through the nearby town of Namie. This is what I found most impressive: it had turned into a contemporary Pompeii—the city had become a ghost town. 20,000 people used to live in it. Now there were only cops guarding all access routes, to prevent anyone from returning home.

We then headed south, stopping for the night in Iwaki, the hometown to Dr. Okada's wife who, out of discretion, had not joined our little expedition. Okada Shinichi, in like manner to Ban Yoshio's associate professor who drove Martine and I in the Eighties around Hokkaido, including sightseeing a perfectly round lake and a sprouting volcano, confined himself, very modestly and in silence most of the time, to his role as a driver.

Since closure of its coal mines, Iwaki has converted into a thermal spa. Shinichi had booked us for the night into an hotel-restaurant, noted for its access to a hot spring. Accordingly, I enjoyed taking a bath in the company of Naoya and Shin-ichi. This was a ritual I already was familiar with, very Japanese: one sits on a small stool, proceeds to scrub meticulously, pouring wooden buckets of hot water over the head and the whole body. A small white towel serves to mop the wet body and for modesty. When one finally enters the hot bath, this towel rests, rather comically, over the head. We then repaired to a private dining room for a delicious meal of half-a-dozen courses.

The following day, after breakfast in the same private room, we hit the road again, along the coast, heading for the Tenshin Museum, perched on a cliff in Ibaraki. Tenshin Okakura (1862–1913) was an artist and gentleman of culture. Through his functions as curator for some years at the Museum of Fine Arts in Boston, he initiated Westerners during the Meiji era and at the turn of the twentieth century into the cultures of the Far East. We then went to see, in Izura, the little studio he had built for himself under his house, in a very picturesque rocky cove. It reminded me of the composer Grieg's, near Bergen, Norway.

We took the road back to Tokyo, stopped for lunch, went on to the potters' village of Kasama—where I found some pretty stoneware to take home as gifts to my wife. That evening, the Nakagawas, Shin-ichi and I had dinner in a tavern near the Makuhari train station. The atmosphere in that inn was happy, relaxed and very young.

On Sunday morning these three Japanese good friends accompanied me to Narita airport where I boarded my return Tokyo-Paris flight. Naoya who hardly talked any more said, "Pierre, I hope you will come back to see us when you are the age I am now (88)." If this were to happen, then he would be 100.

At the end of this latest visit to Nipponia, I was touched once again by the gentleness, the calm and the extreme kindness of the Japanese to me. They were hospitable at all times, delicate and warm-hearted.

A high point of my career was investigating self-assembly of the $5'$-GMP. Take a hard look at the data was one of the lessons of this hunt, spellbinding for a few years that got me hooked. A story was there, waiting to be told. As earlier mentioned, the very first inkling had been a plot of her results by Agnès Paris (b 1950): a straight line alright, but not extrapolating through the origin. It made me curious, I realized it was due to a self-association (Fig. 20.1).

Looking now at the aftermath of this still unraveling piece of biophysics, I am impressed at how important it turned out to be. Just quoting a few publications that postdate my departure from research, "guanine quartets are readily formed by guanine nucleotides and guanine-rich oligonucleotides in the presence of certain monovalent and divalent cations. The quadruplexes composed of these quartets are of interest for their potential roles in vivo, their relatively frequent appearance in oligonucleotides derived from in vitro selection, and their inhibition of template directed RNA polymerization under proposed prebiotic conditions. Stabilization of G-quadruplex structures formed from telomeric DNA, by means of quadruplex-selective ligands, is a means of inhibiting the telomerase enzyme from catalyzing the

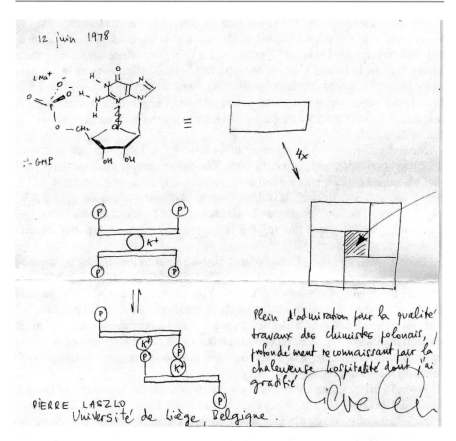

Fig. 20.1 After I made a presentation at the Polish Chemical Society, Warsaw Section June 12, 1978 I was invited to dinner at the house of one of my hosts. Before leaving, I was asked to sign his guestbook. At that time, one might allude to one's work in so doing: my annotation shows my understanding then of the 5'-GMP self-assembly

synthesis of telomeric DNA repeats. The Na$^+$ system is much more regular. High concentrations of K$^+$ shift the equilibrium toward stacked G-quartet formation but different in structure from the antiparallel G-quartet structure of the Na$^+$ system. Each G-quartet is simultaneously stabilized by multiple neighboring interactions. Hence, the addition of a tetramer to a preexisting helix end is favored not only by multiple associations with other monomers from the top tetrad of the helix, but also includes the coordination of the cation, coming along with the G-quartet, with the oxygen atoms of the stacked tetramers. Consequently, the helix chain growth is an energetically favorable, cooperative process. G-quartet structures now surface in areas ranging from structural biology and medicinal chemistry to supramolecular chemistry and nanotechnology."

In short, our work was indeed pioneering and is still being acknowledged.

What was the reception of our own work on clays. It has been most gratifying, to the extent that to this day (2020) our work continues being cited and being influential. I am flabbergasted at its still standing wake. I wrote two review articles on these topics, back in 1986 and 1987, in *Accounts of Chemical Research* and in *Science;* and a book with Balogh Maria (b 1949), 1993. They are not only being cited, being heavily cited: scores of 406, 481 and 431, respectively. However, I resist relishing it, knowing full well that many a scientist will refer to a paper without having read it in full or at all.

More significant is the influence of our work, in Liège predominantly, in spawning studies closely related to ours. We devised novel clay-based reagents that we named *clayfen*, *claycop* and *clayzic*. They have become commercial and in heavy use, still now in 2020. It is interesting to look at use of *claycop*, since 2016 say, i.e., more than 30 years after we had introduced it. The extent to which chemists, for instance in India, use this reagent is impressive, many applications concern synthesis of novel heterocycles.

During my entire career, I worked hard, 70-hour weeks. Now, having been retired awhile, on the eve of 2021, they have dwindled to 20-30 h.

Back to Christian Detellier (b 1952?) who has settled in Canada, more precisely in a French-speaking enclave within Ontario, bordering Québec "*la belle Province.* " I am proud of Christian Detellier and I am glad that he continued—although in his own original way—the work on clays from our Liège lab. He was among my early coworkers in Liège. He and I worked on a number of projects—perhaps most intensely on the $5'$-GMP self-assembly.

His own career (completely on his own) took place at the University of Ottawa. Howard Alper (b 1941) a professor there had asked me to recommend a coworker of mine for a postdoctoral stay in his lab. I complied. Christian's stay was a success, to such an extent that he became a professor at the University of Ottawa, where he ended as vice-president for research—smart decisions on their part.

Christian expressed his gratitude for the training he had had in Liège by espousing in his own research many of the topics from my lab. Even though he did not pursue biophysical chemistry he and I had immersed ourselves in, he most notably carved for himself a niche—original, creative and most useful. He was impressed by the potential of clays in chemistry, that my group in Liège had started exploring. Christian, though, chose not to compete with us and rather find an avenue of his own. We had been working on montmorillonites, so-called 2:1 clays.

Christian chose instead to study kaolinites, 1:1 clays. They are the most abundantly mined clays: to the extent of about 40 Mt. per year. They are layered minerals, the layer associates a tetrahedral silica sheet and an octahedral alumina sheet. Christian exploited the potential of kaolinite as a slow-releasing agent for drugs. He also intercalated polymers to make intercalated nanocomposites. Grafting organic groups onto the aluminol internal sheet led to applications in catalysis, in sensing, in heavy metals adsorption, in exfoliated nanocomposites, in luminescence, and in structural modifications to form nanoscrolls or nanorolls. Christian published nearly 200 papers and they received about 4500 citations, for a most respectable h-index of 39. Not only respectable, highly commendable for a high-level university administrator as Christian was.

John Goldsmith teaches linguistics at the University of Chicago. He is also one of the best educators I have ever come across. Recently, he wrote about dynasties of charismatic teachers, referring explicitly to Brentano, both Freud and Husserl were students of his. In John's case, it was Noam Chomsky. Charismatic teachers are identified by having started dynasties of similarly gifted people.

I have no idea whether I qualify as a charismatic teacher. I can only observe that supervising a research group is also teaching—arguably of the best kind, one-on-one. I feel honored having had students of the caliber of Christian Detellier. I should mention also in that vein Lionel Delaude, Armel Stockis, Michel Moutschen or Onaka Makoto. Of course, my role was to pass on what I had been taught by persons such as Mademoiselle Guillet, Messieurs Guittard, Jourdan, Duny, Bahiana, Francès, Grua, Deluchat, Deleuze, ...

I was only a link in a chain: as the song goes, a success in France by Mouloudji (1922–1994) in the Fifties, *tu n'es qu'un maillon dans une chaîne, Tu n 'es qu'un moment de la vie. Un moment de joie, de misère. Et puis on t'enterre.* (you are only a link in a chain. You are a mere moment in life. A moment of joy and of misery. And then you get buried). / Sénergues, Christmas Day, 2020.

Personal Papers and Public Archives

21

Creativity and invention? Many experiments by psychologists show that a disorderly work environment can be favorable.

Inventory? Just consider the etymology of the word: from Old French *inventoire* "detailed list of goods, a catalogue" (fifteenth century, Modern French *inventaire*), from Medieval Latin *inventorium*, alteration of Late Latin *inventarium* "list of what is found," from Latin *inventus*, past participle of *invenire* "to find, discover, ascertain" (see *invention*).

This chapter deals with the transfer of documents from the personal, private and, generally speaking, uninteresting; to the public domain, in the frozen publication form of archives—public archives. How and why did I become involved in such activities? What kind of documents are worthy of conservation? How do they speak to the historian? I shall strive, from experiences I have had, to answer these queries.

I remember my very first visit to a public archive. This was in Grenoble, I was about 12 and a teacher took the class to the *Archives Départementales*, where one of the archivists made a presentation. I recall a single point, but vividly: he proudly showed us the essay Henri Beyle—best known through his *nom-de-plume*, Stendhal—had written for his *baccalauréat* exam. This—the fact that such a document had been saved—made quite an impression on me.

Fast forward to the present day: I have donated my personal documents for archiving to the main library at *École polytechnique*, in Palaiseau; What does this donation/deposit consist of? About 55 handwritten notebooks, that I kept between 1979 and 2002. These were commonplace books, in which I wrote down what was interesting to me: contents of the seminars I attended; drafts of articles; points to make in presentations; titles of books for reading or purchase; ideas for new research; poems of mine; names and addresses of people I had met; etc.

How is a sense of history acquired? How does one become an historian? What qualifies as an historical document? I shall try to answer these inter-related questions through my own example, since this is an autobiography.

The first inklings came from my father when I was a child. He would return to Grenoble from business trips to Paris with old books which, often, he would give

© The Author(s), under exclusive license to Springer Nature Switzerland AG 2021   181
P. Laszlo, *A Life and Career in Chemistry*,
https://doi.org/10.1007/978-3-030-82393-1_21

me: nineteenth-century scientific textbooks and popularizations. He knew I liked such presents because, whenever I was sick and he had to leave me at home by myself, I would read avidly similar books that Madame Laronde, the librarian at his Neyrpic workplace, would select for me.

These old books—about a century-old—were engaging, if not downright fascinating. Their musty smell and yellowing pages gave me a feeling, not only of time elapsed, of time erased, too. I would thus submit that a taste for history comes from the fear of dying—in my case, clearly linked to my mother's death when I was not quite 7.

Old books, OK. But what about letters from one's contemporaries, addresses exchanged with someone, other scraps of paper? What made me save such documents apparently devoid of any future interest or to an historian?

The novels of Jules Verne, for sure. Very often, they start from such a premise. *Les Enfants du Capitaine Grant*, for instance, is a circumnavigation around an entire parallel: children search for their lost seaman father on the basis of a partly destroyed message indicating that it was written *somewhere* on that geographic line.

Here is my contention: a scrap of paper, a piece of waste seemingly, may have great historical value. The historical narrative—history in short—consists in running a connecting thread through a (finite) number of otherwise disconnected paper scraps or artefacts. To me there is an analogy of the historical record with this familiar sight on a beach, as successive assaults by wavelets leave an ephemeral imprint on the wet sand. Many an object has interest to the historian, it thus resembles leashes from a tide.

I shall now turn to some examples and to documenting some of my researches in history of chemistry. I became acquainted with and addicted to archival work through university libraries, first and foremost the Firestone at Princeton, dating back to the fall of 1962. Others, quite a few others, were to follow.

For instance, because the memory has etched itself into my mind, I used in 1981 the reading room of the Morgan Library. It is located in Midtown Manhattan, on Madison Avenue. This was before the building renovation by Renzo Piano. The books I studied there? Lamarck's *Philosophie zoologique*; Wordsworth's *Descriptive sketches in verse*; Goethe's *Essai sur la métamorphose des plantes*; Shelley's *History of a six-weeks tour*. What I have not forgotten either is that the neighbor, to my left, was working on the autograph manuscript of a Haydn symphony. A few weeks later, I discovered the equally superb and memorable Beinecke Library, at Yale in New Haven, where I studied the global circumnavigation (1815–1818) by Adelbert von Chamisso (1781–1838) on the brig "Rurik." Chamisso was both a botanist of the first rank and a superb writer.

I have become used to the little routines of archival work. You arrive at an usually modern building. You check your belongings into a locker, bringing in with you only your notebook and a pencil. The librarian receptionist assigns you an empty desk in the reading room, after you have handed in a card spelling out the desired item. It arrives after a while, in the form of a cardboard box with an inventory and a number of folders, with in it originals, such as letters sent and received, schedules, all sorts of records.

For instance, I was in Aubière, on the campus of the University of Clermont-Ferrand—more precisely in the building of the Institute of Chemistry. I was inspecting their most disorganized archives. I came upon a letter, mailed out in 1933 by a former Head of the Institute, Professor Léonce Bert. This was addressed to the parents of the 20 or so students in the senior year. Bert was informing them of the itinerary and cost of the study trip he had organized, to allow these students visit some of the industrial plants in Southeastern France, in Saint-Etienne, Lyon, Marseille and Grasse.

I was able, using that letter as my point of departure, to reconstitute that trip in some detail and to publish a whole article about this, so to say, radiograph of an important part of the French chemical industry in the Thirties, that centered on soap, cosmetics and perfumes—the latter a specialty of Léonce Bert (to whom I shall return). Had this document been discarded, I would not have been able to benefit from a whole trove of highly useful information.

Others have described the same eerie feeling. As one scours moldy archives, as a rule, only distant murmurs emanate from them. Very rarely do they erupt into life. At such privileged times, one can hear a voice from the past. An individual with joys and torments springs out of the boxed papers.

A few years ago, I had such an experience. I was thus perusing archives in Strasbourg for another project. They recorded the 1939 transfer of the University of Strasbourg to Clermont-Ferrand, in Central France, where the faculty and students would be compelled to remain for the entire duration of World War II.

One of the files was labelled "L'Affaire Bert" (the Bert Case). As I read it I became transfixed. Academic quarrels are fierce internecine wars. Moreover, the one I was glimpsing resulted in the expulsion of one of the main protagonists from his academic heaven. That all of this occurred during the German Occupation of France turned it into a tragedy. Anti-Semitism was, if not a component, at least a factor.

I was finally able to piece together the story after also mining archives in Clermont-Ferrand, both those of the Puy-de-Dôme *département* and of the former Chemistry Institute of the University. In so doing, I could not help becoming fascinated with Professor Bert.

His was a rich personality. He was troublesome, yet benevolent. He was both a self-made man and at the same time a pure product of the French educational system. He was an innovator scientifically, but a conservative, even a reactionary research worker. A born educator, he was also a political activist. A Socialist at first, he had gradually drifted to the Right, to such an extent that in 1940–1941, he held Vichy-like views.

He left precious few personal documents. Moreover, resentment towards their father led his children to dispose of most of what he had bequeathed. Aroused by the scarcity of documentation, I felt challenged to better uncover his life.

His was the biography of a totally forgotten scientist, even though he was active not long ago and at the forefront of French science. To tell his story provided me with another challenge—or set of challenges.

Can one reconstruct the life of a near-anonymous scientist in like manner to Alain Corbin's feat (in *Le monde retrouvé de Louis-François Pinagot. Sur les traces d'un inconnu, 1798–1876*) of narrating the life of a person chosen at random? In spite of the distrust of the genre by professional historians could I reconstitute a useful biography? One that might contribute to history—of science, of the French meritocracy, of the Twenties and the Thirties, of the French chemical industry, of the dip in quality of French science during the interwar period, of perfume chemistry, . . .?

Léonce Bert was a chemist, and I am a French chemist as well. Hence I can claim some insight into his work, his publications especially. Had I known him, he would have been the equivalent of my grandfather, scientifically speaking. Yet, we came from the same mold. The highly conservative French educational system underwent little change between his experience and my own. It valued the selectivity and the prestige of its education programs for teachers. Thus, I know at first-hand Bert's formative years. It gave me a rather intimate knowledge of his intellectual make-up, an explosive mixture. He was a rebellious individualist who found himself in a straitjacket. He had ample time for donning this accoutrement, a quarter of a century of training, from kindergarten to his doctorate, not to mention further regimentation during the Great War.

Our two careers, his and mine, could not be identical. Time made the difference. In the aftermath of, firstly the Dreyfus Case and secondly the Great War, some of France's leading scientists became politically involved. Bert was part of that group and he took to the political pulpit as a Socialist. My generation was almost the polar opposite of his, at least in that respect. We had a deep distrust of politics.

He and I both chose the current "sexy" field within chemistry. Bert opted for perfume chemistry in 1920. I opted for physical organic chemistry in 1960. These two sub-disciplines differed markedly in their goals and tools. I realized that Bert's field demanded rehabilitation, having fallen in the dustbin of history, for reasons which I shall come to. Suffice to say that the launching in 1920 of Chanel's N° 5 perfume, an economic and a social event, was also a landmark in science and technology.

We are fortunate that Léonce Bert's major misfortune made him in succession an academic chemist and later an industrial chemist. His life thus illustrates the two faces of chemistry, as both a science and an industry. The book will document their differences and their interplay.

Bert's switch from a scientist to an engineer-inventor occurred as he was reaching 50. Chemistry differs from mathematics and physics, in that the latter reward youth. Conversely, chemical research is a slow maturation process. Bert had still much to offer that the abrupt change prevented him from expressing. That is the main reason why he did not become famous and for history of science to have passed him by. Which is not to say that study of his scientific interests, of his contributions, of his research and writing styles, of the institutions he was associated with, does not yield a cornucopia of both questions and answers.

The year 1922, when Léonce Bert obtained his *agrégation* degree, was still in the shadow of the Great War. The French were bent on revenge, crippling and humiliating Germany. The British, less adamant, agreed to decrease the amount of German reparations in exchange for their guaranteeing the Versailles Treaty, which led to the resignation of Aristide Briand, the French Premier. Raymond Poincaré, an advocate of an intransigent policy towards Germany, replaced him. French Germanophobia was also fueled by the use of chemical weapons in 1915 by Germany. Fritz Haber led the German effort in the production of toxic gases. The Nobel committee nevertheless chose him for its 1918 prize in chemistry, rewarding the Haber Process for fixation of atmospheric nitrogen. But the French regarded him as a war criminal. The two countries, France, and the Germany of the Weimar Republic, suffered from the same plight, inflation. In France, prices increased threefold after 1914, and doubled again between 1922 and 1928.

It was still an era of colonial empires. France's was huge, extending over several continents. Indeed, a Colonial Exhibition opened in Marseille in mid-April 1922. Politically, an event which would affect Léonce Bert, was the conference held in Tours at the end of 1920. It signaled the split between the future French Communist Party, subordinated to Moscow, and the SFIO, a more moderate, more democratic brand of socialism. Bert would eventually become an active member of SFIO. The sciences saw the public vindication and acclaim of Einstein's relativity theory with Eddington's astronomical measurements during a solar eclipse. It was the heyday of atomic physics, Niels Bohr received the 1922 Nobel prize in physics for his planetary model of the atom while Francis W. Aston received the chemistry prize that same year for devising the mass spectrograph. The American Gilbert N. Lewis in 1916 advanced a theory equating bonds between atoms to paired electrons. He would go on to evolve a theory of acid-base behavior also based on electron pairs, published in 1923. The First Solvay Congress in chemistry was held in 1922. Among the French participants of this high-level meeting were several scientists who would become important to Léonce Bert in later years, Marcel Delépine, Albin Haller, Charles Moureu, and Georges Urbain. They were respected and elderly senior scientists, members of the French Academy of Sciences.

As a young scientist Léonce Bert strove to publish in the Academy proceedings, *Comptes-Rendus de l'Académie des Sciences*. To do so, it was necessary for any communication to be approved by a member: this was part of the French all-powerful patronage system, also of its gerontocracy. The above-named French senior chemists had also participated during the war in the French program to produce lethal gasses, the French military did not lag far behind Germany in resorting to chemical weapons.

During the early Twenties, the French writer Emile Meyerson wrote landmark works in philosophy of science with theories of a decidedly anti-positivist bend, as supported by conservation laws which he construed as fundamental. His stand as a realist led him to emphasize scientific explanations, overcoming mere descriptions. In a not-unrelated effort, Hélène Metzger, who had also begun as a philosopher of science, redirected herself toward the history of chemistry. During the war she wrote *La genèse de la science des cristaux*, that she used in 1918 for her D. Sc. dissertation. In 1922, she put the finishing touches to a magnum opus, published the following year, *Les doctrines chimiques en France du début du XVIIᵉ à la fin du XVIIIᵉ siècle*.

At the turn of the twentieth century, French meritocracy contended with the harsh reality of an increasingly tenuous and difficult livelihood in agriculture. The population in the countryside, 8.25 million in 1900, dwindled to 6.45 million by 1930.

Mountain agriculture suffered even more. The demographic pressure for mere survival on farms with an already small acreage, together with health improvements, forced many of the children out. In an area similar to that in which Léonce Bert spent his childhood, the Champsaur-Valgaudemar, exodus to the cities affected 7.3% of the population over a 10-year period. Intellectually gifted children escaped into the teaching profession, with the village schoolteacher encouraging them to do so, which applied to Léonce Bert. The French meritocratic ideals offered him a way out of brutal rural existence, together with a rise in society, a guaranteed income and retirement pension, along with access to culture—an all-important feature of French life. He benefited from scholarships, enabling him to receive an education at the secondary and university levels, with virtually no cost to his family. This was a time, the 1890s and the early 1900s, when French public education was still marked in depth by the recent reforms of Jules Ferry. The French revered a school degree, whether the *baccalauréat*, the *agrégation,* or access into one of the *Grandes Écoles*, it guaranteed entitlement to a better life. A prime example of French meritocratic advancement was that of Alfred Kastler, Nobel prizewinner in physics and Léonce Bert's near-contemporary. His brilliant career illustrates French Republican egalitarianism and school system at their best.

At the beginning of the second decade of the twentieth century, when Léonce Bert was studying for his *licence* (BS), French science was triumphant. Becquerel and the Curies had won the 1903 Nobel prize in physics for their discovery of radioactivity. The Nobel prize in chemistry went to Paul Sabatier in 1912, for his work on heterogeneous catalysis, jointly with Victor Grignard, for his work on magnesium organometallics. This event, much celebrated in France, was a factor in Léonce Bert's choice of chemistry as his field of study. Other contemporary developments in chemistry relevant to the story, were the worldwide hegemony of the French perfume industry, based in the town of Grasse, near the French Riviera, where chemists made use of extracts from locally grown flowers, together with imports from the French colonial empire; the rise of the chemical factories of Poulenc Frères in Lyon; and, more distant, Paul Ehrlich's invention of chemotherapy and his highly controversial devising of *salvarsan*, a drug effective against syphilis. In 1912 Bert, then 20, began study in two separate universities, Lyon and Dijon, for degrees in physics and in chemistry, respectively. Why the hurry? While he had the ability to strive for these degrees simultaneously, his family means were too modest to see him through university for more than a year or two. He succeeded in both endeavors during the second half of 1913. Bert was encouraged by his professors to try for the *agrégation* the following, the fateful year 1914. He succeeded at the agrégation in physics, together with some illustrious names of twentieth century French science. Why has this all but disappeared? I turn now to his downfall, which I was also able to discover in the archives.

During the interwar period, in the late 1930s, war again loomed and seemed inevitable. As a border city, the French government foresaw the capture of Strasbourg by the Germans, and took steps for that contingency. In particular, the University of Strasbourg, because it would be a major prize, was to be transferred west. Clermont-Ferrand was chosen for the relocation.

When war indeed broke out in 1939, the plan was followed. The entire University of Strasbourg, its equipment, faculty, staff and students, moved to Clermont. Both universities, Strasbourg and Clermont, were to share the same buildings, although each was to retain its separate identity and administration.

Accordingly, Alsatian chemists were housed in the chemistry institute headed by Professor Bert. He proved to be a warm, congenial and welcoming host as they settled into their new, cramped quarters, during the fall and winter of 1939.

Afterwards, there occurred a series of incidents, minor brushes at first, between the Alsatian chemists and Léonce Bert. It was fully chronicled, it was part of Monsieur Bert's temperament not only to resent challenges to his directorial functions, but also to ask his superiors for support. In this way, he cloaked his authority in theirs. Given his personal ambivalence concerning any authority overriding his own, he could not refrain from a sarcastic tone in such dealings.

Also on account of all the reports he was handling as an elected official, including complaints from citizens, Léonce Bert had developed a habit of going through channels in writing. Such behavior, however precious to the historian scanning the archives, does not always ingratiate the complainer in an academic environment, as compared to a city administrator. Even university administrators, deans for example, do not take kindly to being treated as cogs in a machine or as bureaucratic public servants. They prefer to be seen as independent scholars who, out of sheer altruism and public spiritedness, take on unrewarding tasks for the good of the community. They tend to resent being called upon to referee feuds between colleagues, always rampant but sometimes raging. They hate having to handle formal complaints. Their talent is in subtly conveying hints, to resolve conflicts informally, without stirring the pot. They prefer to remain friends with everyone under their jurisdiction.

When I came upon these documents and was able to piece together the files from the public (*Départementales*) archives in Strasbourg and Clermont-Ferrand, the whole unfolding story was crystallized. It greatly moved me. Others have already told of the unique feeling when suddenly uncovering from dusty archives the authentic voice of individuals from the past. Thus, I followed the self-inflicted downfall of Léonce Bert. It was pathetic, a true tragedy worthy of the Greek dramatists.

(It might be a good idea to merely transcribe for this planned book, in that particular chapter, in chronological order the series of administrative reports to the Dean of the Faculty of Sciences and to the University President from both sides, Strasbourg and Clermont. They make for eloquent reading, as one watches the unfolding of this modern Greek tragedy.)

Bert could not hide his spots, he was despotic, even petty at times. The first conflicts were trivial. The Alsatian scientists were dedicated research workers, 24/7. Bert would lock the gate to the *Institut de Chimie* at nights and on weekends. The Alsatians colleagues failed to respect the closed gate and opened it repeatedly.

He was, he wrote, and it makes for uneasy reading since the Vichy ideology shines through, a man of order. He was training young people for work in industry, he wanted them to stick to regular hours and to be disciplined in every way. Unbeknownst to him, though, his immediate superior, the Dean, Professor

Emmanuel Dubois, had joined the Resistance.

Dubois must have been rankled by Bert's ramblings, especially given the personal attacks in print waged by Bert just a few years earlier due to political differences. Their antagonism, Bert a *pétainiste* and Dubois a *gaulliste*, reflected the division of the French in two camps and was the context for what followed during the spring and summer of 1941.

During the spring of 1941, Léonce Bert noticed that the supply of benzene, all of 20 liters of benzene was missing from his Institute. It was a dramatic loss. One cannot overestimate the importance of this solvent in an institute of chemistry. It is used in teaching, for student laboratories. It is used in research as one of the main solvents required to run chemical reactions. Moreover, benzene is a key solvent for a perfume chemist such as Léonce Bert.

I have to use some metaphors to communicate the importance of benzene solvent to a chemical laboratory. Not having it is like a household without electricity, an airplane with no kerosene, a baker with no yeast. One cannot do without it.

Dr. Bert was indignant. Not only was he aggrieved, the climate of relations with the Alsatians had soured by then to the extent that Bert was quick to accuse them of the disappearance. They angrily denied having had anything to with it. Accusations flew back and forth.

Dr. Bert wanted his benzene back. He wanted at least to find a culprit. He was outraged. He notified the academic authorities, but they were incapable of helping.

In despair, Léonce Bert filed a formal complaint against the Alsatians. In response, the Alsatians filed a complaint against Bert.

The police came to investigate and they quickly found the guilty party, Pierre Bert, Léonce's younger son. The eldest, Paul, was then a student in *Institut de chimie* and Pierre was also frequently hanging around the lab. As the above metaphors hinted, benzene, besides being a solvent, is also a hydrocarbon fuel and Pierre had stolen the benzene to run his motorcycle.

At that point, the official inquiry was closed. Pierre Bert was only 16, he was a minor and he could not be prosecuted.

Léonce Bert had no choice but to do, formally, the honorable thing and cover up for his son. He said that *he* had taken the benzene, needing it for his work. See Pierre Laszlo, "The University of Strasbourg and World Wars," in *Sciences in the Universities of Europe, Nineteenth and Twentieth Centuries,* Boston Studies in the History and Philosophy of Science, vol 309, Springer, Dordrecht, 2015, pp. 89–105.

This was the proverbial last straw. It destroyed utterly Professor Bert. Its president had had enough, the Dean of the Faculty of Sciences had had enough. The University of Clermont-Ferrand in October 1941 fired him from *all* his academic positions. 21 years of hard work, the most advanced degrees in French education had gone up in smoke, puttered away by the tailpipe of Pierre Bert's motorbike. The only concession the academic administration made to Léonce Bert in getting rid of him was a promise to keep the whole rigmarole under wraps and to hush it up.

There circulated two versions afterwards. The Bert family somehow convinced themselves that Paul Pouchet-Gayet, the mayor, together with Léonce Bert, had patriotically refused to turn over the keys of the city to the Nazis. The dates do not jibe—no matter. The personnel of the *Institut de chimie*, colleagues in the University, were told that Bert had been ousted because he was—a widespread accusation at the time, the excuse for many other inequities as well—a Jew and a Freemason (the latter he may have been).

Léonce Bert's expulsion from academic paradise was as extraordinary as it was sudden. That it was hushed up is unsurprising. Most university professors, with the exception of Léonce Bert, utterly dislike and carefully avoid any public turmoil.

Lecturing Trips

22

The notion of a worldwide scientific community to some extent is a pleasing exaggeration. Colleagues one becomes acquainted with appear to share similar values as yours, whether political, intellectual or cultural, whatever their country or language. To some extent, this is an illusion, one tends to project onto others one's own values.

With that caveat, the scientific community is an endearing reality. To be able to travel far and wide. To reach a foreign country which can be extremely exotic and be nevertheless hosted, perhaps within a family home, by someone sharing your values, who in like manner to yourself, may have other interests besides science—such as the outdoors, literature, art or music, can be delightful.

All of that makes it worthwhile to take lengthy flights, to undergo jet lag, not to mention indignities airlines often indulge in—such as delayed flights, mediocre in-flight meals and the discomfort of seats in the economy class.

Interaction with colleagues whom one thus starts to befriend occurs through a number of channels, such as journal articles dealing with topics resembling yours, the attendant peer reviews, meetings at conferences, whether in a lecture room or, more importantly, for meals together. Not to forget, newsletters that many a field of science sports and, this wonderful habit given us by the Americans—accustomed to long-distance interactions between family and friends—the end-of-year newsletter.

Obviously, interacting with other members of the worldwide scientific community is only occasional. It happens neither on a daily nor on a weekly basis—it is all the more precious for its relative rarity. I shall recollect in this chapter a few such occurrences that were memorable for one reason or another.

Unexpectedly, during my first trip to Australia I acquired a lifelong friend—not so much within the scientific community, but within that of the community of scholars. An intellectual historian, Jamie Kassler (née Towle Croy), the wife of Michael Kassler, who after graduate studies at Princeton became a musicologist and computer consultant specialized in robotics. Jamie, also American, is a musicologist, who has published extensively, both articles and books predominantly about the remarkable English writer and polymath Roger North (1651–1734).

I was directed to Jamie by an historian of science, specialized in Darwin, whom I had met at one of the universities in Sydney, David R. Oldroyd (1936–2014). Jamie and Michael own a beautiful house and garden in the Northbridge district of Sydney, thus named for its proximity to the famous suspension bridge. They have hosted me in their home, which I found to be a real privilege, in part because of the book treasures they own—more on that below. At the time when I was preparing my book, *Citrus*, for publication they photographed me in front of their orange tree, a shot I am very fond of. Their garden was not large, but it was paradisical. I vividly recall hearing, every morning, the liquid melodious song of the kookaburra bird, it was astounding. Jamie and Michael have donated their papers—6 boxes of manuscript material, correspondence mostly +1 med folio box +3 map folios, all in all measuring 2.5 m of shelves—to the National Library of Australia.

Jamie visited us and stayed in our house in Plainevaux, near Liège. She befriended Aline, the youngest of my three children. At the time, Aline who was born in 1977, was 10 or 11. Jamie invited her to stay with them in Australia. Most unfortunately since Aline would have loved it, this not come to pass.

I have a number of other recollections from various stays in Sydney, including a disquieting memory: I was once a guest by myself at dinner in a private home. The lady of the house was a most unpleasant person. She was an astrologist who published in Australian daily papers and ladies magazines. She asked me the birthdate of Martine, my then wife. Stupidly, I gave her the information. She then informed me, most assertively, that Martine was in imminent danger, she was undergoing some severe health problem. Of course, I did not believe it but, nevertheless, called Belgium as soon as I got back to my hotel room, to ascertain the total absence of a problem.

I have quite a number of other and warmer recollections from my stays in Sydney. The noteworthy arrivals at the airport, you had to pass in front of a bunch of most athletic-looking young men, they were in charge of detecting drug addicts! The French people working at the Alliance Française. Likewise, an evening in the company of the remarkable historian of chemistry and science popularizer, who doubles up as a gifted actor, Daniel Raichvarg. Plus of course, walking along Botany Bay, where Captain Cook first landed on the new continent.

Arrival in New Zealand by air has always charmed me. Was it the first time or subsequently? In any case, I had a window seat to savor the beauty and the greenery on the hills, as the plane flew in final approach to the Auckland airport. Moreover, the person on the adjoining seat, I found out from chatting briefly with her, was an authentic Maori princess.

I loved New Zealand, on every single visit there, the very first in 1979. A major surprise was thus becoming reacquainted with England, as I had known that country in the years immediately following the end of World War II: the frugality of New Zealanders, their homeliness with a distinct flavor of the Thirties; their familiarity and ease with the outdoors—whenever they have a few days to escape on their sailboats, they roam the Pacific; Allan Odell, my host in Auckland, was a personal friend of Sir Edmund Hillary (1919–2008). On my first visit, Allan, upon hearing that one of my first scientific papers concerned the structure of rimuene, drove me to a nearby beach to show me the tall rimu trees which, in the nineteenth century, served as masts for the clippers seagoing ships.

There were peculiarities, that to me characterized New Zealand. One was crossing streets in Auckland at that time. There were street lights to allow the crossing upon flashing green. But it was not the usual simple matter of green or red: in addition to crosswalks at right angles to one another, there were also diagonal crosswalks, which a set of streetlights also incorporated!

To mention another, one evening during a party of colleagues, the wife of a chemistry professor gave me her marmalade recipe: for years, I made delightful use of it in our family; and reproduced it in my *Citrus* book.

A memorable side-trip was the week I spent skiing on a still active volcano, Mt. Ruapehu. It is nearly 3000 m (10,000 ft) high and thus becomes snow-covered in winter. In the center of the North island, it is about 300 km distant from Auckland. I travelled by bus in order to reach it. What was most memorable about my stay? The absence, at that time, of ski lifts. One would clamber inside the trailer portion of a SnoCat, going moving on tracks and be thus carried toward the top of the mountain. Plus, the frozen lava, in sheets or pillars, covered with ample snow, made the views spectacular and out of this world.

While at the University of Liège, I had to travel often to the United States. For the Brussels-New York crossing, I would get on a Sabena flight, the Belgian airline. Detractors claimed this acronym stood for "Such a bloody experience, never again" at best an exaggeration.

The airline flew Boeing 747 s. One would board the flight at JFK in New York, in mid-evening, and it would arrive in Brussels the following morning.

I was once seated, in economy class, on the right of the cabin, with a middle-aged American couple seated next to me. We started chatting. They were Mennonite missionaries returning to their African posting, after visiting families and friends in the Midwest. I had never heard of Mennonites, they explained to me the origins of that Protestant sect, who was Mennon and how some ended up in the US. Mennonites are pacifists, one of their most endearing features. These two persons were kind, they were congenial and very pleasant to be with. Hence, after dinner had been served and cleared, we whiled away the time with occasional talk. The hours went by.

After a long time, they told me that they had revealed a lot about themselves and their religion. But what about me and my religion? I answered that I was a Catholic atheist. That statement flabbergasted them. They asked me, most politely, to explain that oxymoronic statement. I went into how I had been raised a Catholic and as the years went by had lost the faith. But nevertheless still felt a part of the Catholic community. I felt concerned, for instance, by what the Pope was doing and saying. Was it Roncalli, John XXIIIrd, at that time? Perhaps. In any case, they gulped hard at my declaration, took it in and remained thoughtful and silent for a long time.

Meanwhile, our flight was proceeding. We were in-between Iceland and Ireland when I saw, to the left side of the plane, the beginning of sunrise—an event I love watching from a plane. I got up and walked over to the tiny port in a door, and just stood there for some time admiring the dawn from about 30,000 ft. up.

When I resumed my seat, the Mennonite lady had a large smile and she said to me, "there are some displays in nature, wouldn't you think, Sir, that one cannot help ascribing to God?" Wickedly, I answered this assertion of her's with "Beauty is in the eye of the beholder, Madam." As far as I recall, this was the end of my theological discussions with this lovely couple.

Fort Collins, where Colorado State University (CSU) is located, flanks the Rockies, about 100 km (60 mi) north of Denver, on the rectilinear Interstate road heading north to Cheyenne. I visited there several times after having flown into Denver.

My host there was Gary E. Maciel (1935–2014). After his Ph. D. with John S. Waugh at MIT, he spent most of his career at CSU, after a number of years on California campuses. I believe he had joined CSU university from his love of the outdoors. He was an outstanding nmr spectroscopist. He was also the first American I knew who drove a pickup truck. It turned in time into a fashionable trend. Once, Gary and his wife Maxine drove with me into Denver and we had drinks together on top of a tall building there, enjoying the view. It may have been by occasion of the seminar I presented there on July 22nd, 1983.

Anyway, on one of my several visits to CSU and to Gary, he invited me to join him and a few graduate students for an outing at lunchtime. We drove up the La Poudre Canyon—the name was given it by French explorers of yesteryear (Cache La Poudre = Hide the Powder).

And we went rafting downriver. There were three or four separate rafts and 12 or 15 of us. Gary and I were the only faculty, all the rest were graduate students in their early 20s. Hence, it was a boisterous party. Some of the students fell into the water but clambered back into a raft. There were rapids—actually the stream consisted almost exclusively of a succession of rapids, for around 10 miles or so. We must have been on the river for in-between two and three hours. I found it to be highly enjoyable—more than a retribution/honorarium for the lecture I had given, a prize truly.

Where was I invited to give lectures, when I was heading research groups at the University of Liège and the *École polytechnique*, thus being based in Western Europe, in both Belgium and France? To the United States, predominantly. It was of course necessary to spend many hours on flights.

And how did I spend the time on such flights? In work and in leisure. In work, writing publications from our research group. It first entailed also analysing the numerical data, for which a calculator—the HP-25 at first—came in handy. In leisure, reading scientific reprints and literary books.

In addition I listened to music on all those long flights. In 1981, when this kind of apparatus was still brand-new, during one of my visits to Wesleyan University—in Middletown, Connecticut—Phyllys Rose (b 1942) sold me on the Sony Walkman. It had just come out worldwide. She said that the music seemed to come from inside one's own body. Indeed, after I had purchased one, to me it was a novel magic.

Thus, I would pack a number of recordings, around a dozen, in my hand luggage. At that time, this was before CDs, they were small rectangular cassettes, about the size of a pack of cigarettes, only half as thick.

What would I listen to, then? Bach music, first and foremost, whether organ pieces played by the likes of Marie-Claire Alain or Helmut Walcha; keyboard pieces, played on the piano or the harpsichord, by Glenn Gould in particular, the Goldberg Variations. Quite a few other soloists and composers: for instance, Emil Gilels in the N° 2 Concerto by Saint-Saens; Mussorgski's "Pictures at an Exhibition," in either the piano or the Ravel-orchestrated version; of Ravel's, the Sonatine; Bartok's "Toccata for Two Pianos and percussion," played by Georg Solti and Murray Perahia; and so on, and so forth; songs of Georges Brassens's too.

Sometimes, infrequently, I would strike a conversation with someone else on the flight. It might be a fellow-passenger I sat next to. Once or twice, on board a 747 jumbo, I spoke at some length with the pilot, who had left the plane to the copilot and was eager for conversation—I thus recall the chief instructor pilot of the Air France company, a very interesting man.

And once, this was when we lived in Pinehurst, North Carolina, and American Airlines offered a direct flight from Paris Orly to Raleigh-Durham, North Carolina, I struck a conversation with an air attendant, Jacobi Daley-Vanderkooy. She regularly flew that route and I met her again a couple times on these self-same flights. I told her about the splendid system of hiking trails—so-called GRs—one finds in France. She told me of a recent walk she and her husband Jerry had recently taken near Mont Saint-Michel. They enjoyed many other hikes in France. At the end of one of these flights into RDU, I introduced her to my wife Valerie who had come to fetch me. And the two of them became good friends.

What do I like most about Japan and about Tokyo? The feel of Asia in the little streets, with the *noren,* little banners hanging over the doors of shops. The magnificent stationary stores, such as Itoya in the Ginza district. Millions using public transportation, thus making for orderly traffic.

Primarily, reverence by many for the past and continuity with previous generations—the cult of the Emperor being one of its manifestations. It does not interfere with a vigorous hold on the present.

Like France, Japan is ruled by hierarchies. The pecking order among universities sets Tokyo in first place, with Kyoto as number two. Hence for an academic, getting one's toe in that country demands doing so at the top. I gave a lecture at the University of Tokyo on October first 1985, I believe. I was invited there by Oki Michinori (1928–2016), whom I had known at Princeton, when he was spending a sabbatical in Kurt Mislow's research group.

In the aftermath of the lecture, I was taken for dinner to a nearby Japanese inn—sitting on the floor at a low table. There were about 16 of us, and Nakagawa Naoya (b 1927)—who was hosting me in his house—seated next to me. I was the only foreigner in that group.

Japanese colleagues, whenever they got together, were like a bunch of schoolboys, wanting to have fun. In Japan, it is customary at a get-together of fellow-workers in the evening to direct the talk at denigrating their boss. I witnessed it on that evening.

At that time in the mid-Eighties, Japan was not yet a very open country, nor a destination favored by many tourists. Japanese knew foreigners from their interaction with Americans mostly. And they knew well the American reticence to foods strange to them, such as raw seafood. They were keen on testing their foreign guest

Set to have some fun at my expense, the group started by ordering sushi for everyone. I love sushi, I gobbled it up. They then ordered sashimi. I love sashimi too that I also devoured.

Then, they giggled a bit and ordered *natto*. This was a dish I had heard about but never been confronted with. It consists of fermented beans. It looks and smells awful—just like what you have in mind.

The waiter put down in front of me a large plate, heaped with the stuff. I dug in: I am a child of the war, or rather of wartime, and I am used to eat whatever is served on my plate. After some time—and I must confess, it was not easy—I cleared the helping of *natto*.

At that point, my friend Naoya nudged me and said: "do you know what you have done, you just went through the order of *natto* for the whole table."

I attended a conference at the University of Stirling, in Scotland—a country that justifiably prides itself on no fewer than 15 universities for a population of only about five million. My recollection is of a small and verdant quiet campus. As is generally the case with conferences held in British universities, we the participants were housed on-campus, in a facility for students. Stirling is a small town, located on the railroad between Edinburgh to the East and Glasgow to the West. The conference lasted for a week.

On the second or third day, or rather on that evening, I repaired to the room of a fellow participant, a Brazilian named Roberto Rittner Neto (b 1941), who came from the University of Campinas, in the State of São Paulo. He had suggested that I come around, so that we would become better acquainted. It was a joy to meet, I had spent happy years in Brazil as a teenager during the early 50s.

Did we become buddies! Even though there were no alcoholic drinks that I recall, Roberto provided music from a tape recorder, I do not recall him having brought a guitar. In any case, I was thrilled to meet a Brazilian with whom I could share my memories of his wonderful country and culture. For hours, he and I belted out together popular songs and successes from past Carnivals: "*Maria Candelaria/é alta functionaria—trabalha trabalha/na letra O–O–O*" and many others, including *Sasaricando*. This started a friendship that has lasted to this day.

This recollection from Stirling is accompanied by two or three others from Edinburgh, where I stayed for a few days after the Stirling event. One such recollection is of having joined another two French participants from Stirling, Claude Coupry (b 1948) and Christian Brevard (b 1944), both of whom also became good friends over the years. The three of us found ourselves, one evening, on the Edinburgh Castle, seated on bleachers, trying to stay warm—an occasional gulp of a local product was very helpful—and watching a Royal Tattoo—the name for a uniquely British military parade.

Another fond memory of this particular time in Edinburgh, where and when I must have also met with my sister Ilona and her husband Hugh Morison, who live there, is of having attended the final concert of the International Harpsichord

Competition. It was presented, not by the prizewinners but by their examiners, the members of the jury: in the spirit of British fair play at its best and indeed it was outstanding.

Claude Coupry and Christian Brevard both would go on to most accomplished careers. Claude, at the time of our meeting in Scotland, was using Raman spectroscopy expertly; and to great avail, when she turned it upon archeological and historical artifacts. She became an authority on ancient pigments and their compositions. She also enrolled in the team studying frescoes in the medieval abbey Saint-Germain in Auxerre.

Christian Brevard early on became the head of the Bruker Spectrospin company manufacturing nmr spectrometers in Wissembourg—a lovely ancient little town in the Alsace, next to the German border. As a fringe activity, he became a banker—or rather was picked for the board of directors of a large French cooperative savings institution.

The Laszlos dwelled in Rio de Janeiro during the period 1950–1953. After we returned to France, I remained nostalgic of Brazil. For one thing, my teacher of chemistry, Senhor Bahiana, had decisively pushed me in that direction.

Brazil, however, remained an elusive destination, throughout the military dictatorship that muzzled it from the Sixties until 1985. In 1986, thus at the first opportunity, I flew into Rio to attend an nmr meeting there. The suggestion had come from Roberto Rittner, whom I had first met as mentioned in Stirling, Scotland.

The conference was held at the Hotel Gloria, nested at half-height on the Morro de Gloria hill, overlooking the bay of Rio, across from the Sugar Loaf. At the first coffee break, I found myself engaged in a shouting match with another participant, older than myself and much more knowledgeable about Brazil than I was. I was naively telling Brazilians my joy at being back and telling them how I had impatiently awaited the end of military rule, having vowed not to set foot in Brazil as long as the military remained in power.

At that point, John B. (Jake) Stothers (1931–2012), a Canadian, stepped in and most forcefully argued the opposite. Many among the Brazilians present supported him: he had had the merit, during these difficult times, of coming often to Brazil and of converting many to modern nmr methodology. He had been their lone contact with the science world at large. After we had both expressed our positions, our clash came to a stop. We did not have nor seek the opportunity to reach afterwards some sort of resolution.

Other recollections of this 1986 short stay in Rio? One is foremost. I got on the ancient streetcar climbing—or rather escalading—to the Santa Teresa district. There I discovered, never having been there before, the Chácara do Céu mansion and park. It offers a superb view over the city and the bay, and is surrounded by this most picturesque ancient neighborhood of Santa Teresa.

However short—a couple of weeks—this 1986 stay in Brazil convinced me of the plight of underdevelopment. I met quite a few young Brazilian colleagues: as a rule, they had handsome training, honed by a Ph. D. and as often as not a postdoctoral stay in top American universities. As soon as they came back to Brazil, though, they became sterilized, scientifically-speaking, engulfed that they were, as starting university teachers, in 30 or even 35-contact hours weeks, teaching students; thus making it impossible for these superbly trained and beautifully gifted young people to continue doing science. A shame.

What were the contents of this traveling lecturer's bag? I must have been representative of the species, as it roamed the wealthiest parts of the planet from the Sixties until the new century.

In addition to my clothes, toiletry bag, shaving utensils and aftershave, comb, a small manicure set; of course two or three sets of slides, already on their trays, in these bygone time of Kodak carousel projectors; a pointer, prior to the laser variety, one of my coworkers had presented me with a converted old fishing rod, it was about a foot long at rest and three fully extended. A bunch of music cassettes for listening on my Walkman, starting in the Eighties. I had in my wallet a card with phone numbers all over the world for use to call home. Likewise, a couple of credit cards. The medications I needed during the trip; in addition, perhaps some Imodium for turista.

Plus, of course, gifts to bring to my various hosts. In East Asia, they were prescribed and almost ritualistic, one would get them at the airports in duty-free shops: brandy bottles, but only of canonical brands, Courvoisier and Martell—*armagnac*, for instance however valuable to a Frenchman, was out of contention.

I had devised an elegant solution to the weight problem: there was in Liège a tiny store on rue Charles-Magnette, next to my Voyages Parfait travel agency, catered first by Madame Coeurderoi, afterwards by Monsieur Gauthoye. I was able to buy, throughout the Seventies, handmade Bruges lace. Invariably, the ladies of the households I had been invited to, for dinner usually, would be both surprised and delighted by such a present.

By 1993, my activity as a science popularizer was public knowledge. One of the most delightful aspects was an invitation to travel to Stockholm in December 1993 at the time of the Nobel awards ceremony. This was not, thus, a trip of mine occasioned by a lecture.

The invitation came from the then Ciba Foundation in London. It owned a lovely house at 41 Portland Place, within a stone throw of the BBC headquarters. I had stayed there numerous times while going through London. For a number of years— until affluent professors of medicine from all over the world raised the ante and secured the place for themselves—it had offered affordable luxury accommodation in Central London.

The trip to Stockholm was the idea of Derek J. Chadwick (b 1948), who then directed the Foundation. We were a group of about a dozen, most of whom were active in science popularization. Of those I recall there were Sir George Porter (1920–2002), who had gained a Nobel prize in chemistry (1967), Alun Anderson (b 1948) who edited at the time *New Scientist*, Jonathan Piel (b 1938), then editor of *Scientific American*, Etienne-Émile Baulieu (b 1926), professor at Collège de France, an expert on steroids who had devised a morning after pill and Stephen Jay Gould (1941–2002). From the accompanying staff of the Foundation, I remember the husband of a lady editor, a military historian by inclination and training hired as a strategist by one of the British chains of supermarkets, Sainsbury's I believe.

After flying-in from our various departure points, Valerie and I arrived in Stockholm, possibly during the weekend of December 4–5. The Ciba Foundation treated us like royalty and put us up at the Grand Hotel, where the Nobel laureates also stayed. It was that difficult time in Sweden having daylight from only about 10 in the morning to 3 or so in the afternoon.

I have recollections from some of the companions in our small group. Sir George, always benign and smiling. Baulieu, making a big show of the faxes he was getting and sending. Gould, intent upon remaining unapproachable.

And I remember of course the set-up for the prize awards ceremony. Dr. Chadwick had managed to get our group on the Thursday at a large table in the Stockholm Concert Hall, where the ceremony would take place the following evening. We were thus privy to the dress rehearsal. At one point, Toni Morrison, who had won the literature prize that year, stopped by our table and chatted briefly. She was resplendent in an evening gown. And we witnessed an endearing and eerie silent parade of young ladies, all in white, in anticipation of the Feast of St Lucia on December 13—a Swedish festival to celebrate the forthcoming reappearance of daylight —, a traditional event that many a Nobel prizewinner had noted with awe. The embodiment of Lucia carries a crown of alight electric candles in a wreath on her head. Each of her handmaidens carries a candle, too. It is an outwordly sight.

Reading the previous paragraphs, the mention for instance of a harpsichord concert, you may become envious of us scientists and wonder how we ever get any work done. We work hard, believe me, foreign trips are an integral part of our agendas. That they offer such compensations—and one has sometimes to look hard for them—is, truly, a gift from the Gods.

Now to the trip I took to Norway during the third week of May 1980. It came at the invitation of Johannes Dale (1923–2003), an outstanding scientist and a most remarkable person. I had met him initially from our joint careers in Belgium: Johannes had been part of the remarkable group of scientists gathered, in Brussels, at the laboratories of European Research Associates, an emanation of Union Carbide. Our attendances of the stereochemistry Bürgenstock conferences often coincided. Johannes, after ERA had been dissolved, resumed his career, as a professor at the University of Oslo; which he was when he invited me to a small lecture tour in his country.

I am as fond of Norway as I am of New Zealand, for similar reasons: sparsely populated countries with huge coastlines and tall mountains, people with a rugged spirit and a simple lifestyle, very democratic political system. This 1980 trip of mine to Norway was at a time when the standard of living had not yet been boosted sky-high by hydrocarbons.

A splendid recollection is of my having reached Bergen from Oslo by train: a seven hour ride, but in a luxury setting and going through mouth-dropping landscapes: the railroad crosses the craggy Hardangervidda plateau at a height of 1237 m (4,058 ft) before dropping to its seashore arrival in Bergen. As luck has it, my stay in Bergen coincided with the opening of the Music Festival. No, I was not in the concert hall for the ritual playing of Grieg's N° 2 piano concert. But I was among the few onlookers when the Crown Prince, all by himself and democratically driving his car, got there to open the Festival.

My next and last stop was in Trondheim, where Jostein Krane, a former coworker of Johannes Dale and of Frank Anet, was my host. After my lecture, I was invited for dinner at the Kranes. I vividly recall Mrs. Krane recounting her ordeal. She had gone to Moscow as a student. Was it for a few months or an entire year? In any case, the KGB tried to recruit her as a spy for the USSR. First, threats and imprisonment, with the Gulag lurking. Second, an attempt at seduction, with a handsome young military type escorting her to a show at the Bolshoi. But she had stood her ground and had been released back to her country.

Now to a trip I took in 2003 to the Swedish Polar Station. The plane having left Stockholm flies over countless frozen lakes. The Swedish countryside is dotted with ponds. Some areas show trails, for skiing or snowmobiling. A full moon is on, still visible at midday. In this immense country, the habitat is scattered. Our flight reaches the coast of the Gulf of Bothnia. It is jagged, fractalized. Islets are fringed with ice. The ocean remains free though. Nevertheless, here and there it shows the beginning of icing-up. We reach the other side of the gulf, and fly over Umeå.

Forest now covers the ground. Countless roads can be seen, probably used for logging. Rivers descend from the mountains, I am told they were used to float tree trunks. Now, after an hour's flight, rivers and lakes of snow and ice show total hold of the landscape. The Boeing 327 begins its descent. Quite high mountains are visible in the distance. We get close to a military base with its strips for fighter jets. Just before landing in Kiruna, after an hour and a half of flight, I admire in the ambient white clumps of trees, they show an ineffable softness. And one sees, just south of Kiruna, the huge iron mine. The baggage claim room is decorated with a huge stuffed bear, and with Sami art.

Almost exactly a century ago, Swedes built a railroad to carry iron ore into Narvik, a Norwegian harbor about 180 km distant. They built the extractive infrastructure and the city started from scratch is now home to 50,000 people. The municipality of Kiruna is the largest in Europe. Its 20,000 km^2 might accommodate the entire population of the planet. Later on, a road, the E10, was opened parallel to the railway, along the southern shore of Lake Toreträsk, a subarctic Lake Geneva with a depth of 300–400 m. At the same time, at the beginning of the twentieth century, the Swedish Academy of Sciences established a science station, now also a nature reserve, at Abisko, by the lake, about 130 km from Kiruna, at 68 $^\circ$ 21 'N latitude. This is where our meeting would be held.

It is very, very beautiful. The landscape is serene, even more than noble. It is covered, up to about 400 m elevation, by dwarf birch trees, not taller than two or three meters. Their resistance to the relentless turmoil commands admiration. The wind blows at 20–30 km/h. It snowed on the second day, gusting to over 100 km / h. Bending over was necessary to negotiate the twenty meters from the main building to one of the two buildings housing our meeting rooms.

Huge freight trucks roll down the road. At the edge of the lake, trailers station on parking lots, sheltering Lapps. They fish in the lake, from within small tin or wooden huts, through a hole. You can sometimes see people busy themselves nearby. The houses of the village remind me by their colors, ox blood, yellow, ... of the isbas in Irkutsk. Most often they are made of wood. Cars and just as often snowmobiles are parked next door. Stores lack storefronts. Most of the houses are closed in on themselves, shutters closed.

A herd of reindeers crosses the road, the Lapp shepherd follows them on his snowmobile. I see a team of dogs. Two moose come near the research station.

Anchorage, Alaska, was a refueling stop on the way to or from Japan. It was exciting to me, because of my boyhood in Grenoble—a town nested in the French Alps. It left me a lifelong love of the mountains.

The route from Western Europe to Alaska, or vice-versa, came close to one of the most beautiful mountains I ever saw and I rejoiced in the prospect of that view: Mount McKinley, now renamed Denali.

It towers at some 6194 meters (20,310 ft) and is a splendor. After trekking through the Hoggar mountains in the Sahara desert, I had befriended our guide, Albert Coccoz, at the time one of the world's leading climbers. However, a year or two later, he and his wife were killed by an avalanche. Albert had had a wish, to climb and ski down Mount McKinley. It was fulfilled by our mutual friends, Odette Bernezat (b 1940) and Jean-Louis Bernezat (b 1936), also experienced mountaineers.

Whenever the flight I was on skirted Mount McKinley, I was glued to the window.

Trips to Poland were easy when I worked at the University of Liège: this city is on the railroad going from Paris to Moscow through the then two Germanies and Poland. I may have visited Poland half-a-dozen times. Some trips of mine happened during the Communist era, I also made more trips after 1989.

A recollection from the former: colleagues drive me in front of a large building in Warsaw. They tell me: "these are the headquarters of our friends." And then proceed to explain: "the enemies of our enemies are our friends." (it was the Chinese embassy, at a time when the USSR and China did not see eye to eye). They also told me the terms of the economic exchanges between Poles and Russians: "we give them our wheat and in exchange they take our coal."

Views that became familiar were of the central squares in cities, Warsaw being one of them, reconstructed in the identical style they showed prior to World War II: they gave me a Potemkin-like impression of a cardboard background. But I loved the countryside, it reminded me of the French countryside in my childhood. Apples tasted the same and were likewise wormy, I just loved them—the apples, not the worms.

Some Polish colleagues were distinctly weird. For instance, while in a biomolecular institute, where research on nucleotides carried on, talking with one of the senior scientists there, I put to him the question of the origin of life—to me very much legitimate given his area of research. He retorted with this non-answer, "I believe in God, Sir, and to me that is the end of it." Such an aggressive non-answer froze my heart.

While on the topic of awful Polish encounters, I have of course to recall my visit to Auschwitz, by occasion of a stay in Cracow. My main impression was of the smallness of the place and of my incredulity at the sheer contrast between the size of the buildings and the awesome numbers of their all-too-temporary forced occupants.

In June 1978, I went through Wroclaw (the former German Breslau). In the evening, I had a ticket—these were very scarce and in great demand, my hosts had the foresight to purchase mine in advance—to a show staged by Jerzy Grotowski (1933–1999), a luminary of the theater in the twentieth century. This was "Apocalipsis cum figuris." It was memorable, a major theatrical emotion: religious ceremony-like; spectators in a small number, around 30 only; Bunuel-like, a succession of surrealistic scenes; body-language primacy, with the actors mistreating themselves; the very first scene was a masturbation. We, the voyeuristic onlookers, were uncomfortably seated on the floor, our backs against a wall.

Another and also memorable evening watching a play, during one of my Polish trips, was with Tadeusz Kantor's (1915–1990) "Dead Classroom." He entrusted his ideas to amateur actors only. After it was over, I found myself attending a party and chatting for at least an hour with two of these actors. The apartment we were in was rather opulent, which surprised me, given the dire state of the Polish economy. It belonged to a couple, both welcoming and forthcoming. They were artists and produced for connoisseurs in Italy. Such an outlet gave them a very comfortable life. Polish expediency at its best!

In 1996 I was bedridden with a severe low back pain and sciatica for several months. What sustained me and got me through was the prospect of going to . . . New Caledonia. For work, not leisure: I was booked for a week at the Orstom Institute in Nouméa, to train some of its leading scientists, a dozen, in science communication. Orstom—it has been since renamed IRD—is a French governmental agency doing scientific research (oceanography, climatology and meteorology, geology, anthropology, etc.) in former French colonies, predominantly.

I was lucky, in my ailing condition, that the Paris desk of Orstom (Madame Catherine Hartmann) accepted to foot the expense of the plane tickets for me and my wife in business class. As it turned out, we travelled in the small upper salon of an Air France 747.

This is a long journey: New Caledonia is almost at the antipodes of Western Europe. It takes about 24 hours to reach and the time difference, depending on the season, is 9 or 10 hours.

Everything about that week in Nouméa turned out to be eerie. We were to get there by flying into Tokyo, refuel and go on to Nouméa. But there was a typhoon. We could not land in Tokyo, we were told, and would be diverted to Osaka. At the last minute though, this was rescinded. Our plane was the first allowed into Tokyo-Narita. The headwind was still so fierce that the plane stopped on the runway in record time. The terminals were filled with people lying down everywhere and having done so for the long hours during which the hurricane had raged and stopped all airport traffic.

The second leg of the long journey, took about eight hours. It was eerie, I tell you: because after all that flying, after having travelled all the way from France—we arrived in France: a cartoon-like France, with not only French speakers, but beret-wearing and *mobylette*- or 2 CV-riding Frenchmen and French bakeries. As Valerie and I were driven from the airport into town in the early morning, the weather was splendid and we could see quite a few young Japanese tourists in pairs, honeymooning couples, New Caledonia was a popular destination in Japan for the newly-married.

We were driven to the Orstom center in Nouméa. It was a large prefabricated building dating back to World War II, Americans had built it as the headquarters in the General MacArthur-led reconquest of the Southeastern Pacific from Japanese forces. Valerie and I were given use of a small apartment in that complex, where my lectures would also take place.

It was a small classroom, with a desk and a projection screen above and behind it. Most of the time, my back condition forced me to lie supine on the desk. Nevertheless, the lecture series went very well. The audience—I remember the names of Yves du Penhoat and Joël Picaut, oceanographers both—was extremely friendly and cooperative.

Our stay had a Sunday in it. The then director of the Orstom Center, François Jarrige (b 1939) took us sightseeing over the island in his car. He was driving recklessly fast on the mountainous roads, with many hairpins and turns and, usually, little visibility. Some of the stops he had planned were indeed of great natural beauty.

We were invited several times for dinner at their house by Paméla and Yves du Penhoat, they were highly congenial and could not not have been friendlier and more hospitable.

It was a short walk from the Orstom down to the beach and the lagoon: quite a sight, with bathers, water-skiers, kiteboarders and the ever present Japanese tourists—their country had lost the war militarily but had won the ensuing peace, economically. Once, Valerie and I witnessed a bizarre scene—our stay was strange, I tell you—a group of 15 or so young Japanese together with us on a sandbar, the water depth was at most 5 or 6 ft., and they were sporting, for photographing or filming one another of course, full scuba-diving paraphernalia, complete with wet suits and oxygen tanks.

The lagoon waters were divine in temperature and dangerous because of some of the creatures they harbored: tiny *aiguillettes* fish darting every which way, and the *poisson-pierre*, masquerading on the bottom as a stone but with a nasty, venomous sting. The locals get into the lagoon as a consequence only when well-shod with *claquettes* sandals, also known there as *tabis*.

Unfortunately, we did not get to interact with the indigenous Melanesian population, the equivalent to New Zealand Maoris or Australian Aborigines, the Kanaks. The one meeting place, we were told, between Europeans and Kanaks, were the kava bars, where many people—men predominantly—repaired to at the end of the day. Kava is a traditional Melanesian drink made from the root of *P. methysticum*, a relative of the pepper tree. Kava has a relaxing effect. It is found throughout the Pacific Ocean cultures and is mainly consumed in the Melanesian islands: New Caledonia, Vanuatu, Fiji and Solomon Islands. We met with Pierre Cabalion, an ethno-pharmacologist whom Pierre Potier had recommended to me. He was an expert on kava and was enthusiastic about its drinking.

Books That Were Influential

23

To paraphrase the Haeckel (1834–1919) dictum, 'ontogeny recapitulates phylogeny', the books I read in English and French were instrumental in my development as a scientist and as a person. Thus, this chapter presents some important books to me. Here I reminisce about my nonspecialized scientific reading. I was lucky, for the start of my career around 1960 to coincide with a Golden Age for science popularization. Sputnik, the advent of space exploration and the rise of research universities all whetted the appetite of the public for science stories of any kind. For instance, *Scientific American* was thus reborn in 1948, by 1984, circulation had grown 15-fold.

Some leading scientists used this opportunity to reflect aloud on issues—philosophical issues—such as ethics of science, creativity in art and science, the nature and context of discovery, the march of science in its chaotic state, with its slowdowns, accelerations or brutal swerves. Thus, I shall comment here on some among these books that were important to my development as a scientist.

But first a word about celebrity scientists. Prior to World War II, there were only a handful: Galileo, Newton, Lavoisier, Darwin and Einstein. In the aftermath of the war, accrued many more, due to nuclear weapons, rocketry, electronics and computers, antibiotics, plastics, ... One could reel off many names in these and other areas, of Nobel prizewinners in particular. The media, radio and television in the 60s, the Internet beginning in the 90s, put quite a few scientists into the limelight—not to mention scientists claiming center stage for themselves.

Following the recommendation of a physicist friend, in 1968 I purchased the three volumes of Richard P. Feynman's (1918–1988) *Lectures on Physics*. I loved reading them. The third one, on quantum mechanics, was tougher going (Fig. 23.1).

This was during my time at Princeton. I was then attending informal joint seminars between chemists at Princeton and at Rutgers (in New Brunswick), that Paul Schleyer (1930–2014) at Princeton and Ed Wasserman (b 1932) at Rutgers had set-up. Ed Wasserman had chosen volume number 3 as the textbook for an advanced graduate course he was then teaching at Rutgers.

© The Author(s), under exclusive license to Springer Nature Switzerland AG 2021
P. Laszlo, *A Life and Career in Chemistry*,
https://doi.org/10.1007/978-3-030-82393-1_23

Fig. 23.1 Cover of
Feynman's *Lectures on
Physics* (author's library)

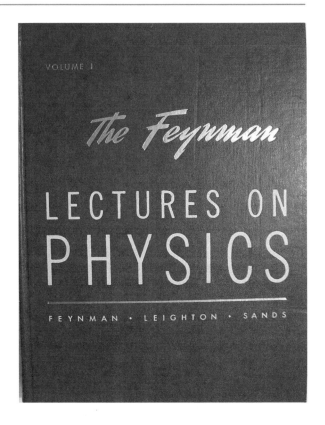

The *Lectures on Physics* to me definitely were master books. They influenced me
to entitle—was I pretentious—my own textbook on chemistry, for beginning
students, *Leçons de chimie* (I owe them my appointment at *École polytechnique*,
12 years later). Hermann, the Parisian publisher owned and headed by Pierre Berès
(1913–2008), brought out *Leçons de chimie*, also in three volumes, in 1974. The
first, entitled *La liaison chimique,* was influenced in turn by another master book,
Linus Pauling's (1901–1994) *The Nature of the Chemical Bond* (1939). Pauling's
book originated in his Baker Lectures at Cornell. Later on, at the turn of the present
new century, I had the honor to lecture as a visiting professor—which I recount in
another chapter—in the same historical auditorium at Cornell. Moreover, earlier on,
when visiting the Pauling archive in Corvallis, Oregon, I held in my hands Pauling's
copy of *The Nature of the Chemical Bond*, in its 1967 last edition—with the revised
title *The Chemical Bond*: Pauling had very carefully prepared the next edition which,
had it seen the light of day would have been posthumous. He had inserted at the
relevant places slips of paper bearing the additions he had prepared. There were
dozens of those. Their tenor? Factual.

I applied and was accepted into Charles A. Coulson's summer school in theoretical chemistry, held in Oxford at the Mathematical Institute in the summer of 1961. I
purchased a copy of his handsome book, *Valence*, a new edition of which had just

Fig. 23.2 Two friends, Jean Massoulié and I in 1994 (with permission from Laurent Massoulié)

been brought out. Thus, I was introduced to molecular orbital (MO) theory, which I found to be as useful as it was wieldy. When, during the ensuing 1962–1963 period I was a postdoc at Princeton with Paul Schleyer, I was a little shocked to find Americans using the rival Pauling valence bond theory. But it was rather easy to learn and emulate. Nevertheless, I remained loyal to MO theory, assisted in so doing by another two outstanding books, Andrew Streitwieser's (b 1927) *Molecular Orbital Theory for Organic Chemists* (1962) and John D. Roberts's (1918–2016) *Notes on Molecular Orbital Calculations* (1961).

When it came out in 1968, James B. (Jim) Watson's (b 1928) *The Double Helix* was a best-seller. It presented science as a contest and the two main protagonists, Francis Crick (1916–2004) and Jim Watson as, at best, shady characters. An untypical gesture on my part for such a mass phenomenon, I bought the hardback as it came out and read it, one evening, in a single sitting.

My main reason for doing so, in retrospect, the purchase and the avid reading both, was my proximity in the chemistry department at Princeton, to Jacques R. Fresco's (b 1928) lab. Jacques was the exact contemporary of Jim Watson. Jacques's lab had been among the first, chronologically, in this then brand-new and glamorous field of molecular biology. My close friend and classmate from NSE, Jean Massoulié (1938–2011) had served a postdoctoral stay in Fresco's lab, in 1960–1962 (Fig. 23.2).

A secondary reason, at least to me, was the association of Jim Watson with the Cold Spring Harbor Laboratory, illustrious for the biology meetings held there—where the frisbee was widely held to have been devised.

As is well known, *The Double Helix* tells of the race to elucidate the DNA structure and to beat Linus Pauling to the finish line (Pauling yet again). The author, a brash young American whose unflattering self-portrait—perhaps reflecting childish insecurity—may be taken as representative of the post-Sputnik generation: aggressive and cynical, opportunistic and self-centered, self-promoting (a Steve Jobs (1955–2011) for instance emulated Jim Watson in such traits). Plus, as my readers surely are aware, *The Double Helix* shows Watson and Crick as unfair to Rosalind Franklin (1920–1955): misogynistic and paternalistic, their relationship to her and her critically important work was inelegant at best.

Science in *The Double Helix* is summarily depicted as competition—rather than as a quest for knowledge and the truth. Winning is all-important: a quote contemporary with the elucidation of the DNA structure as a double helix is from a then coach at UCLA, Russell (Red) Sanders (1905–1958): "winning isn't everything; it's the only thing."

I vividly recall a lecture at Princeton by Paul Schleyer (1930–2014) upon his return from his first trip and lecture tour in Japan. He was very impressed by the Japanese engagement with science, the dedication, the hard work, the masters and their followers—nevertheless he predicted that American science would prevail due to a foremost trait—this ability at competition and at winning it.

Such lessons—admittedly crude and gross simplifications—were not lost on Bruno Latour (b 1947) and Steve Woolgar (b 1950) at the time when they wrote *Laboratory Life* (1979). They encouraged sociologists of science, Latour himself and a rapidly growing band of followers, self-styled post-modernists, to depict science as a power struggle, a cartoon commensurate with Thomas S. Kuhn's (1922–1996) description of scientific revolutions as the overthrow of a previously dominant paradigm.

The resounding public success of *The Double Helix* was due to its description of the scientific discovery as accruing to flamboyant scientists, steeping at nothing in their hunt for the big prize, not averse at cadging data from Franklin and Wilkins (1916–2004), and beating to it more experienced scientists such as Linus Pauling and Erwin Chargaff (1905–2002).

Martine, my first wife, was a medical doctor—a general practitioner. While still at Princeton, I presented her with a subscription to a medical journal: the *New England Journal of Medicine* which enjoyed the top ranking at the time, the end of the 60s. This subscription followed us to Belgium in 1970.

Whenever one of the weekly issues arrived, I would also give it a look and, perhaps, read a few pages. Thus did I become addicted to one of the columns. It was entitled "Notes of a biology watcher" and it was written by Lewis Thomas, MD (1913–1993). At the time, he was at Yale, in the medical school, a professor and a dean.

The Lewis Thomas column consisted of short texts, no more than a couple pages usually. As their overall title implied, they aimed at informing the medical profession of the current status of biology, considered as a whole: from natural history to molecular biology. Dr. Thomas did it in a most attractive way, without any technical jargon even though each essay was well documented, by the sheer strength of an arresting prose and of philosophical considerations. He deemed himself a biology watcher, yet at the same time he was an observer of the human scene as if from Sirius—detached and involved at the same time. These are contradictory stands indeed, but Dr. Thomas managed somehow to span that great distance.

I read these essays admiringly. Which I was prepared to do, earlier articles had sensitized me to the cogency of the short essay. In particular, in my native French language, the *Propos* by the philosopher Alain (1868–1951), published during the first half of the twentieth century.

I allow myself here three rather short quotes from Lewis Thomas, in order to convey some of their flavor:

"Ants are so much like human beings as to be an embarrassment. They farm fungi, raise aphids as livestock, launch armies into war, use chemical sprays to alarm and confuse enemies, capture slaves. The families of weaver ants engage in child labor, holding their larvae like shuttles to spin out the thread that sews the leaves together for their fungus gardens. They exchange information ceaselessly. The do everything but watch television." (July 8 1971)

"Tennis has become more than the national sport; it is a rigorous discipline, a form of collective physiotherapy. Jogging is done by swarms of people, out unto the streets each day in underpants, moving in a stolid sort of rapid trudge, hoping by this means to stay alive. Bicycles are cures. Meditation my be good for the soul, but it is even better for the blood pressure." (December 11 1975)

"We are learning language from each other, so it is environmentally determined, but only in the limited sense that the environment is us. When we are at our real work with each other, talking and writing, making music, painting intensely private versions of the world for everyone to see, are we uncoupled from biologic influences? Or are we, literally, genetically, made this way? Who can say? Well, anyone can say is the answer, because we really do not understand these things, not yet." (June 29 1978)

In the meanwhile, between 1971 and 1978, Dr. Thomas had moved from New Haven, Connecticut, to New York City, having turned from a Dean at Yale to President of Sloan-Kettering in New York—the latter an institute for reaearch and treatment of cancer.

29 of the "Notes from a biology watcher" became gathered into a book, *The Lives of a cell*, in 1973 already. Other collections were to follow and were equally successful.

Those writings by Lewis Thomas had their influence. I became likewise a practitioner of the short essay form: in my monthly "Petite chronique archéologique," in *Nouveau Journal de Chimie*; in a "La nature et l'artifice" series in *La Recherche*, published mostly in 1996 and 1997.

One word, admiration, expresses my feelings from reading Gaston Bachelard (1884–1962).

This occurred during the Sixties. They concerned, on one hand, his writings in philosophy of science published in earlier periods; and, on the other hand, his *Poétique* books which by and large then appeared. A vivid recollection is of my short walk from Lycée Saint-Louis, where I studied in NSE as related in earlier chapters, to the José Corti (1895–1984) bookshop facing the Luxemburg Gardens: Corti doubled up as a publisher who brought out a number of the Bachelard books.

What did I like and admire in Bachelard's impressive output? There was the man and his by then legendary rise from humble beginnings to prestigious academic stature: what is termed in France *l'ascenseur social*, the social elevator. Bachelard had started as a postal delivery man and postal clerk, self-taught himself in maths and philosophy, had served as a secondary schoolteacher and run the gamut of the difficult, competitive French certification exams, culminating in the *agrégation* and doctorate, prior to being appointed to a chair in philosophy at the Sorbonne. An exemplary and absolutely unique and truly awesome trajectory.

His philosophy of science books had appealing traits. Instead of the all-too habitual focus on physics by other philosophers, he was very fond of my chosen field of chemistry; and showed an incisive understanding of it. He was versed in alchemy too which seemed to fulfill a need of his for imaginative texts bridging his two major interests of science and poetry. He also showed an impressive and intuitive sense of science in practice. His pronouncement that science insults commonsense appealed to me for its insight. Reading such a notation of his brought to my mind the canonical example of a quantum particle passing simultaneously though two holes and interfering with itself.

During that same period of the Sixties, I also read Thomas S. Kuhn (1922–1996) *Structure of Scientific Revolutions* (1962): what Kuhn termed revolution and described as a switch in paradigms, Bachelard called *rupture épistémologique*: to me the two notions by the American and French philosophers of science were virtually identical. The explanation was mutual ignorance. French contributions in that time did not register in the Anglo-Saxon world. Then of course occurred the great upheaval when, suddenly, American academia engulfed itself in trendy ideas from French thinkers: structuralism at first, followed by the rise to prominence of Derrida (1930–2004), Foucault (1926–1984), Lacan (1901–1981), Bourdieu (1930–2002), Baudrillard (1929–2007), etc.

Gaston Bachelard's enthusiasm for science was unclouded by concern about possible nefarious fallout from technological applications. To him, science was progress and his view of the history of science was downright whiggish, for instance in *La formation de l'esprit scientifique* (1938). Science had the merit of bringing the mind into previously unsuspected exotic new territories, there was no telling where it would go next. The Bachelardian epistemology was a welcome escape from the Cartesian philosophy of science. His views of science came from an insider, which appealed to me.

And his fresh and lively writing style thrilled this reader.

Thomas S. Kuhn's (1922–1006) *The Structure of Scientific Revolutions* appeared in 1962. Only in November 1969, while still at Princeton but knowing that in January 1970 I would start upon my new appointment at the University of Liège, did I walk to the U-Store (the Princeton University Store, on-campus) and bought myself a copy: doing it in Liège surely would not be as easy. Why did I wait all those years for this purchase? I suspected that the amazing popular success of this book—

one million copies sold, translations into 16 languages—did not make it for me a compulsory read. Only my departure from the US triggered its purchase.

It is one of those books that owe their appeal in no small measure to the opacity of their writing. It is a feature that the reviewer of the *Times Literary Supplement*, the following year (issue of November 20 1970) indeed condemned. It was exemplified by Kuhn's use of the word 'paradigm' whose very obscurity made its fortune—it got used and abused in every field, from psychology to economics.

I did overlap with Kuhn at Princeton. He was a professor in the History Department until 1970, when he moved to Cambridge, Massachusetts and a chair at MIT. While in Princeton, I did not take the opportunity to seek him out for a chat. However, I vividly remember the daily sight of him walking his two magnificent dogs (Afghans, I believe) on Washington Road, past the Frick building of the chemistry department. I had close friends in the history department but the topic of Kuhn's merits were never part of our conversations (Fig. 23.3).

From reading the *Structure*, I got the impression of a dogmatic personality. Hence, I was not surprised by his clash with Karl Popper (1902–1994), another fierce dogmatist. In that very public dispute, reminding me of that between C. P. Snow (1905–1980) and F. R. Leavis (1895–1978) a few years before (from 1959 on)—about the Two Cultures—I took Kuhn's side (privately, I had no reason to enter that quarrel). Indeed, when I read the *Structure* in November–December 1969, I liked very much Kuhn's description of scientific activity: this historian of science had been a physicist and he knew from inside what he was writing about. This was my dominant feeling.

Only years later, did I learn that Kuhn had lifted from Michael Polanyi (1891–1976) his presentation of scientific revolutions—such plagiarism did not surprise me unduly.

However seminal and stimulating, the ideas of Karl Popper were late in entering the French scene and debates. The German original *Logik der Forschung* dates to 1934, its English translation *The Logic of Scientific Discovery* was published in 1959, a French translation was brought out only in 1973; we can thank Jacques Monod (1910–1976) for it.

I was fortunate in having discovered Popper and that book during my second stay at Princeton (1966–1970). What did it leave me with? Predominantly, the criterion of falsifiability as demarcation between science and pseudo-sciences: its pertinence to psychoanalysis was devastating and, since Freudian ideas and practices were then (1960s and 1970s) very much in the ascendant, I was not unhappy to see that pretentious balloon punctured.

Apart from what can only be considered as a fringe benefit, I remained uninfluenced by Popper. However, a book made me a passionate onlooker of the philosophical debate between Sir Karl Popper and Thomas Kuhn. This was *Criticism and the Growth of Knowledge* (1970). It gave me the sick feeling of ganging-up to roughen Kuhn, which I ascribed, at least in part, to envy on the part of the group of three (Popper, Lakatos and Feyerabend) at the popular success of *The Structure of Scientific Revolutions*.

Fig. 23.3 Cover of my copy of Kuhn's *Structure*

I felt very much on Kuhn's side in that debate. What I liked in Lakatos's presentation in that book was his insistence on refutations stemming from fully grown and formulated research programmes. Like most readers, I guess, I was amused if not taken by Feyerabend's "anything goes" viewpoint, which obviously captured the mindset of most scientists intent upon debunking a theory to advance their own. Thus, I read with amusement *Against Method* when it came out in the mid-Seventies.

Regarding history of science, I decidedly belonged with the internalists, seeing the advancement of knowledge primarily as an epistemological pursuit. I took solace in books that bolstered this stand of mine. This was decidedly true of the few written by Peter Medawar (1915–1987), a British immunologist and Nobel prizewinner. They described science as I knew it from my own practice. I hope that titles such as *The Art of the Soluble* (1967), *Advice to a Young Scientist* (1979), *The Limits of Science* (1988), not only are still in print but also continue to be read avidly by active young scientists.

Three to four million Lebanese live in other countries than Lebanon. A first wave of emigrants dates to the end of the nineteenth century. The second wave happened during and after the first World War, due to the severe famine caused by the Ottoman blockade and the establishment of the French Mandate. The third wave was due to the civil war that raged from 1975 until 1990.

During the first half of the twentieth century, Lebanese chose to exile themselves predominantly in the United States and in Brazil. Medawar originated in this Lebanese diaspora, that has given us other eminent scientists such as the American chemists E. J. Corey (b 1928) or Bruce Ganem (b 1944). His early years were spent in Brazil—he was born in Petropolis—, another claim on my sympathy since he and I were about the same age when we both went or returned to Europe for university studies and the beginning of our careers.

I loved the way he wrote, always pithy, cogent and to the point. What did I like about some of Medawar's pronouncements?

I emphatically supported his deep conviction of the joint root of imagination in the poet and the scientist. He emphasized that science is both cumulative and predictive: a good research program takes advantage of recent discoveries and builds upon them for practical applications; prediction is a "property that sets the genuine sciences apart from those that arrogate to themselves the title without really earning it."

I could not agree more with his pronouncement about the relative weights of empirical facts and of imaginative conjectures. As he declared in 1965, "the ballast of factual information, so far from being just about to sink us, is growing daily less. ... In all sciences we are being progressively relieved of the burden of singular facts, the tyranny of the particular. We need no longer record the fall of every apple." P. B. Medawar, *The Art of the Soluble*, (p. 114) London: Methuen, 1967.

I loved the clarity of his thinking, as in his asserting (1976) that "we need not persist long in error if there is a genuine determination to expose our ideas to tough critical analysis. In general, it is illuminating to recognize that such human— sometimes all too human—activities as play, showing off and sexual rivalry are not psychic innovations of mankind, but have deep evolutionary roots." P. B. Medawar, "Does ethology throw any light on human behaviour?", in P.P.G. Babeson & R. A. Hinde (Eds.): *Growing points in ethology*. Cambridge: Cambridge University Press, 1976, 497–506.

He stressed the intrinsic simplicity of scientific ideas: this was to me a great encouragement in the dual roles of scientist and science popularizer that I espoused.

During our time in Princeton at the end of the Sixties, we interacted with some of the residents at the Institute of Advanced Study—as individuals rather than as myths. My wife Martine (b 1938) was a resident at the Princeton Hospital. She looked after art historian Erwin Panofsky (1892–1968) during his final illness. I was thrilled by my several encounters and discussions with Harish Chandra (1923–1983) and André Weil (1906–1998) (please, do not read this as an instance of name-dropping). I also enjoyed making more than casual friendships with two younger mathematicians, Alex Grossmann (1930–2019) and Hervé Jacquet (b 1939).

But my admiration for Freeman Dyson (1923–2020) did not stem from such an encounter. Rather, it came from reading—I was then in Liège—an engrossing book, entitled *The Starship and the Canoe* (1978) by Kenneth Brower (b 1944). It featured Freeman Dyson and his son George Dyson (b 1953), as polar opposites who nevertheless managed to communicate.

When it came out just a year later, I devoured the father's autobiography, *Disturbing the Universe* (1979). He recounted his work for the British Bombers Command during World War II; of his contribution, along with Richard P. Feynman (1918–1988) to the renormalization group, while both taught at Cornell; of operational research and von Neumann automata.

During a decade or two at the end of the century, I subscribed to the *New York Review of Books*. The essays in it by Freeman Dyson were delightful, from both his choice of a topic, original and stimulating as a rule, and his writing style, which his *Scientific American* obituary aptly characterized as "both silky and muscular." Another author has described Dyson as an Emerson for our time.

How best to remember and characterize him? He tackled only the big questions, reminding me of Charles A. Coulson (1910–1974); in addition both these Englishmen looked physically alike.

I can convey the not only rich but enriching flavor of his thinking/writing with the following excepts (from the James Arthur lectures he gave at NYU, 1978):

"I hope with these lectures to hasten the day when eschatology, the study of the end of the universe, will be a respectable scientific discipline and not merely a branch of theology."

"If it turns out that the universe is closed, we shall still have about 10^{10} years to explore the possibility of a technological fix that would burst it open."

"no matter how far we go into the future, there will always be new things happening, new information coming in, new worlds to explore, a constantly expanding domain of life, consciousness, and memory."

"it takes about 10^6 years to evolve a new species, 10^7 years to evolve a genus, 10^8 years to evolve a class, 10^9 years to evolve a phylum, and less than 10^{10} years to evolve all the way from the primeval slime to H. sapiens."

"I have found a universe growing without limit in richness and complexity, a universe of life surviving forever and making itself known to its neighbors across unimaginable gulfs of space and time."

The French language has an expression *fort en maths*. It can be rendered as "maths is his or her forte, (s)he is outstanding at maths." Which indeed points to a cultural obsession with maths. The whole French educational system, at the secondary school level, that of the *lycées*, and at the upper level, that of the *Grandes Écoles*, sets mathematics at the top of the curriculum. This pre-eminence of maths has historical determinants. In 1830, Auguste Comte (1798–1857) started his influential classification of sciences by asserting, *la plus ancienne et la plus parfaite de toutes: la science mathématique* (the most ancient and the most perfect of the lot: mathematical science). To give empirical support to the foregoing, to this day France has collected many Fields medals, the equivalent of Nobel prizes in mathematics: as of 2018, France had collected 13 Fields medals, more than twice as many as the UK (6).

I shall thus consider now two highly influential books by French mathematicians, René Thom (1923–2002) and Benoît Mandelbrot (1924–2010), that met with great public success and that I read when they came out with intense admiration.

René Thom's may have been the first. I read or tried to read his book, *Stabilité structurelle et morphogenèse*, when it came out, to great acclaim, in 1972. At that time, Thom who had received a Fields medal in 1958, was a permanent member of *Institut des Hautes Études Scientifiques*, the French equivalent to the Institute of Advanced Study in Princeton.

His was a magnificent intellect, and impressive culturally too. I did not get to meet Thom in person until March 1983 when we both took part in a meeting, restricted to a relatively small group of 50 participants or so and organized in the lovely Alsatian city of Colmar by the European Science Foundation. Entitled "Identification of Scientific Advancement," it aimed at providing a composite picture of the state of science at that time. I was a delegate from Belgium. Other French participants, in my recollection, were Raymond Boudon (1934–2013), Hubert Curien (1924–2005) and Edmond Malinvaud (1923–2015).

Thom, who attended *École normale supérieure*, chose Henri Cartan (1904–2008) as supervisor of his thesis. Awed by the genius of Alexandre Grothendieck (1928–2014), he opted out of algebraic geometry and sought another avenue, in linguistics in particular. In the words of Mandelbrot, "Thom has of course been critical of Bourbaki, and his vivid love of geometry recommends him to every geometer." Which is how he discovered and refined a novel mathematical description of morphogenesis.

His book was seminal. I cannot claim having read it through and I do not believe many people did. Very soon it nevertheless became famous, due to the British mathematician, Erik Christopher Zeeman (1925–2016), who termed it a mathematical theory of catastrophe.

My own receptivity to the modeling by Thom was due, in part, to my having heard during my schooling in NSE of *caustiques (caustics)* in geometrical optics—cusp-like singularities shown by envelopes of light rays hitting a curved surface. They were a particular instance of the singularities Thom dealt with in his book. I must add that I was made receptive to morphogenesis by my education in the NSE classrooms, combining as they did mathematics, biology, physics and chemistry.

Within very few years, three or five, the highly sophisticated treatise by René Thom on morphogenesis had been further reduced to a buzz phrase, by assimilation with Lorenz's butterfly effect: "does the flap of a butterfly's wings in Brazil set off a tornado in Texas?"

Was it Thom's influence or another source? In any case, going through London in 1976, I picked-up at a used bookshop on Charing Cross Road a copy of D'Arcy Wentworth Thompson's (1860–1948) wonderful and highly readable and stimulating, *On Growth and Form,* first published in 1917.

My very first acquaintance with James Gleick (b 1954) occurred before publication of the two books, *Chaos* (1987) and *Genius* (1992) that made him famous.

This was in 1986 when I was visiting at Cornell. I received a phone call. It was Gleick's, then a journalist at the *New York Times*. He asked me questions about clays and their use(s) in chemistry. He was neither courteous nor impolite, just very direct and insistent. The call lasted at least half-an-hour and I could hear the clattering of his typewriter as he wrote down my answers.

I read *Chaos* as it came out. It was a clear presentation of this multi-pronged topic—I was not a specialist but I had been close enough to both Thom and Prigogine to have at least a feeling on that issue. Gleick's book finished making it more than popular, trendy among both scientists and the general public.

A word then about Ilya Prigogine (1917–2003) since I have mentioned René Thom already. A genius or a charlatan? The verdict is not in yet.

I presented him with an honorary doctorate from the University of Liège, priding myself on having done this before his award of a Nobel prize in chemistry. He also ran admirably the Solvay Institutes of Physics and Chemistry at the Free University in Brussels (ULB). The research he directed there (the Brusselator, as it was named) was important and he knew how to second himself with brilliant coworkers such as Paul Glansdorff (1904–1999) and Grégoire Nicolis (1939–2018). I was impressed also by the series of books he co-edited with Stuart A. Rice (b 1932) of the University of Chicago. This is for the credit side. On the debit side, I can testify that he had the ability to speak nonstop nonsensical gibberish—unrivalled except perhaps by the ageing President Mitterrand (1916–1996) during his visit at *École polytechnique* in 1994, with off-the-cuff remarks during his interaction with us faculty members.

When I arrived in Liège in 1970, Prigogine and his associates at ULB were all excited about the Belusov-Zhabotinslii reaction, noteworthy for its periodicity in time and space. I promptly seized it as a demonstration in my classes. I was teaching, in the biggest auditorium of the university sitting 500 students, a first-year class on chemistry. That reaction lent itself nicely to projection on a screen, thus showing its evolution for the duration of a lecture, as it was undergoing its own internal waves.

I return now to James Gleick and his hefty book, his first, about chaos theory. As showed when he interviewed me, he was very conscientious in doing his homework. Hence, *Chaos*—numbering some 352 pages—was comprehensive and free of errors, even if it offended some mathematicians with its insensitivity to what to them is the essence of maths, proof. What Gleick had achieved was portrayal of a kuhnian paradigmatic moment in the history of science.

Benoît Mandelbrot (1924–2010) was, like René Thom, a mathematician of the first rank. He also had to change course early in his career. In his case, this was due to the hegemony of Bourbaki, that he found unbearable; to such an extent that as a young man, after graduating from *École polytechnique*, he left France for the United States, "because of their (Bourbaki's) stifling influence" (his words). There he found shelter in the IBM Thomas Watson Research Center in Yorktown Heights in the Hudson Valley of the State of New York. His work there pursued some of the inspired intuitions of Henri Poincaré's (1854–1912): "for Bourbaki, Poincaré was the devil incarnate. For students of chaos and fractals, Poincaré is, of course, God on Earth. (...) the recent studies of chaos combine the heritage of Poincaré at his sloppiest and the heritage of an already purified Poincaré." Which is how Mandelbrot discovered the set that bears his name.

What did I love about Mandelbrot's theory of fractals? The notion of a fractal dimension. The rebirth of the question of the exact measure of the length of the coast of Brittany. And of course, like everyone else having accessed one of his books, the wonderful images—fractals caught for the first time a fundamental aspect of nature.

And fractals were useful to my catching an important cause of the catalysis by clays of chemical reactions: the reactants get to meet on the surface of clays platelets which are smooth, of effective dimension 2, as contrasted with higher dimensions in the surrounding volume.

Stephen Jay Gould (1941–2002) and Benoît Mandelbrot were good friends. When did I first read a piece of his and become a fan? In the mid-1970s. I did not subscribe to *Natural History*, in which he had a column in like manner to Lewis Thomas's in the *New England Journal of Medicine*. But I read avidly his collected pieces, purchasing his books as they came out, starting with *Ever Since Darwin* (1977). I have at least a dozen on my bookshelves. One that truly struck a chord was *Wonderful Life* (1989).

Why did I like his writings? I have mentioned in this book my taste for short prose pieces. In addition, while never having been a naturalist myself, natural history had been an important part of my education: in the NSE *classe préparatoire*, Daniel Privault (1908–1991)—whom we nicknamed Luigi for some mysterious reason—in like manner to Gould would feature a biological species, an animal or a plant and describe a specimen showing some idiosyncrasy. Such an approach was also Gould's, to which I was thus made resonant by Privault's earlier teaching in the Fifties.

I liked such a treatment so much that I made it mine. For sure, it encouraged a similar format for my *Petite chronique archéologique*, in *Nouveau Journal de chimie*, and my *La nature et l'artifice* in *La Recherche*. In addition, nowadays I write "Plant of the month" entries and post them on my website.

To return to Gould, I was very much impressed by a paper he wrote jointly with Richard C. Lewontin (b 1929). To identify morphology and function is a frequent delusion. Gould and Lewontin wrote their justly celebrated paper on "The spandrels of San Marco" (1979) to denounce this frequent fallacy in evolutionary arguments. But pseudo-sciences feed on its appeal to common sense. Hence phrenology and its mapping of character from facial features and bumps on the head. It would lead to

Broca's identification of diverse perceptual and cognitive activities with various parts of the brain—which thrives nowadays from positron emission tomography and magnetic resonance imaging.

As I wrote in another chapter, Gould and I were part of the same group invited to Stockholm by occasion of the 1993 Nobel Prize awards—but he did not encourage dialog.

I have saved arguably the best to end this chapter. They are the invaluable books by Edward R. Tufte (b 1942) on graphics. Each is a visual feast. Each is a cultural treasure chest in the information presented from a wide variety of sources. Each is exemplary in its didactic approach, taking the reader by the eyes and leading him/her into becoming utterly convinced of the validity of the viewpoint being presented and thus about to be espoused.

Tufte taught more or less in succession political science, statistics and computer science at Yale University, after a start at the Woodrow Wilson School at Princeton.

The Visual Display of Quantitative Information (1982), like the other titles to be listed here, was published by Graphics Press in Cheshire, Connecticut. It sold for $ 34.00. Howard I. Gralla (b 1946) was the graphic designer and typographer.

Actually, the first book by Tufte was self-published. Tufte financed it by taking out a second mortgage on his home. The book was a commercial success—due to word-and-mouth recommendation among scientists. Henceforth, Professor Tufte moved from political science to information expertise. A gem in this first book was the reprinted *Carte Figurative* by Charles-Joseph Minard (1781–1870) depicting the horrendous plight of Napoleon's *Grande Armée* in its retreat from Russia after the failed invasion in 1812—winter having been its worst enemy, as Hitler (1889–1945) found out with his own disastrous attempt at conquest of Russia. Tufte made a sensation of this feat of graphic design he had rediscovered.

I purchased my copy in 1991 of *Envisioning Information* (1990), from the same publisher—Tufte himself. It is a book about how to escape from the page and use its two-dimensional space to present multi-dimensional information. This second book kept to the format and design of the first, a wise decision. Well-chosen examples of poor graphic design were educative, for instance of the New York-to-New Haven train schedule (on page 104) (Fig. 23.4).

I purchased my copy of *Visual Explanations* (1997) as soon as it came out. It was also a most handsome read and a visual delight. For instance, the reproduction on page 114 of a group of butterfly fish demonstrates the features of quality scientific illustration.

Images in a book echo demonstrations during a lecture in the classroom. After presenting some of the questionable evidence supporting the decision to launch that rocket, Tufte gave as an example of effective demonstration the testimony in Congress by Richard P. Feynman of the brittleness of O-rings placed in ice-water, thus explaining the disaster of the space shuttle Challenger on January 28 1986.

The fourth hardback by Tufte I purchased (in 2006) was *Beautiful Evidence*, published in that year, no less admirable than the above three. In the meanwhile, Tufte published in 2003 a short pamphlet blasting the PowerPoint slides provided by

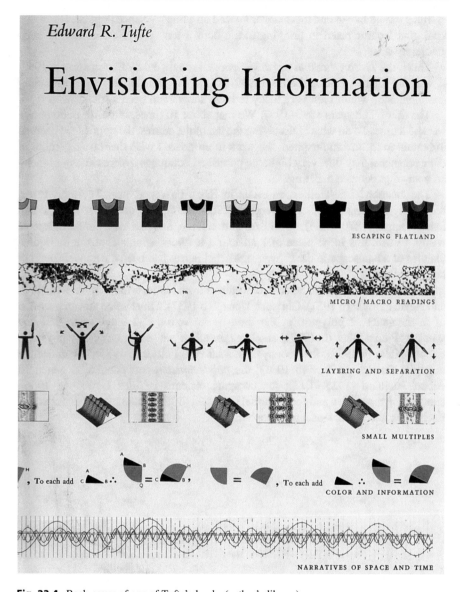

Fig. 23.4 Back cover of one of Tufte's books (author's library)

Microsoft software and he was utterly convincing—I refrained henceforth from making use of so-called bullets in my slideshows.

Nurtured by such masterpieces as I have just related, it would have been a shame if something of their spirit had not infused me. It did! And I have mentioned in passing some of my own attempts at science popularization.

But I should like to end this chapter instead on a book, science popularization of a kind, that I participated in bringing out, a book I am very proud of having been among its writers.

This is the *Trésor, dictionnaire des sciences*, brought out by Flammarion in 1997. It was the idea of Michel Serres (1930–2019), the philosopher and outstanding communicator whom I am very lucky to have known and been a friend with.

The time? The years 1993–1997. We met about 10 times, often for periods of a week at a time. To do what? Discussing the list of the entries, the notions we deemed important to feature and explain, the work in progress. I wish there was a record of our exchanges, they flew very high; the published volume is a mere excerpt of many an hour of passionate debating.

The location? Les Treilles, an estate in Haute-Provence, near Tourtour. It had belonged to Annette Schumberger (1905–1993), a daughter of Conrad Schlumberger (1870–1936), of oil-exploration fame. She had used her considerable wealth to turn this large piece of land into a textbook example of the traditional features of a landscape in the Provence. Michel Serres had been a personal friend of her's.

The team? In addition to Michel, two editors from Flammarion, Sophie Bancquart and Nayla Farouki (b 1951). Christian Houzel (b 1937), a *normalien* and an historian of mathematics, a polyglot, at the time he knew *only* 23 languages, including sanskrit. Pierre Léna (b 1937), another *normalien* and astronomer. As physicist, Etienne Klein (b 1958); the geophysicist Jean-Paul Poirier (b 1935); the molecular biologist Charles Auffray (b 1951); the *polytechnicien* and population geneticist Albert Jacquard (1925–2013); the computer scientists Gilles Dowek (b 1966), another *polytechnicien* and Jean-Gabriel Ganascia (b 1955); and myself as a chemist. Each of us had a streak as a philosopher.

This Was Then **24**

The previous chapter commented on books whose reading was highly influential. Why books? Only books? Why was I so receptive to printed matter?

Why books? From my education: when my parents retrieved me from school on Saturdays at mid-afternoon, they took me to a small bookshop; where they would buy me as a gift the book of my choice, any book. Hence, books became the means by which I built my culture—one at a time.

Only books? Of course not. Newspapers and magazines too. Again, my early childhood was determining. As my Dad came home on his bicycle, I would walk up our street, rue Ponsard (in Grenoble): not only to greet him, also to grab the daily newspaper he carried, *Le Dauphiné Libéré*. I would proceed to read it eagerly while he prepared lunch (it was just the two us, my mother died before I turned 7).

I proceeded to read many newspapers and magazines. Over the years I subscribed to a few: in French, *Le Monde* and its *Sélection hebdomadaire,* when we lived in the US; *Critique—Les Temps Modernes—Le Débat—L'Âge de la science;* in English, at first, the *New Statesman,* the *London Magazine,* the *TLS* and the *New Yorker,* the *Georgia Review, Russell: the journal of the Bertrand Russell Archives,* and the *International New York Times,* after it had replaced the *Herald Tribune.* I was and still am a voracious reader.

What about chemistry? For the duration of my career, from 1963 until the early 2000s, to be accurate, I subscribed to *JACS,* the *Journal of the American Chemical Society.* In addition, I subscribed to *Current Contents,* in its physical sciences edition: it signaled articles in scholarly journals, and I then sent out reprint request cards.

In many fields besides chemistry. I have always considered myself as a generalist, not a specialist. I lived through a time of increased information about the universe: ice from Antarctica informing us on industrial pollution during past centuries—neutrino astronomy—exoplanets—paleontology by the Leakeys—etc.

My luck was in my time the prevalence of the English language in science. Knowing it made it possible to gain information about 85 or 90% of science. I was part of a joint scientific culture, worldwide. One manifestation of it was, during the

© The Author(s), under exclusive license to Springer Nature Switzerland AG 2021
P. Laszlo, *A Life and Career in Chemistry,*
https://doi.org/10.1007/978-3-030-82393-1_24

second half of the twentieth century, the hegemony of two magazines, *Science* and *Nature*, in announcing major breakthroughs.

One recollection to end this segment about what I read: in my late teens, thus in the early Sixties, I met in London—in a tearoom on Leicester Square I believe—a gentleman who started conversing with me. He worked at Rank Xerox, if I recall properly. He was a Marshall McLuhan kind of prophet: he told me of technology making it possible for a single speaker (helped by voice recognition software) to produce an entire newspaper. But authorities—governmental? economic?—had ruled it out since it would extinguish journalists entirely.

My career took place while English was the language of science. My native tongue being French, I had to acquire a command of English—which was greatly helped at first by the numerous times I spent with the Bramwells, in Southwell, England. Later on, living in Princeton, New Jersey, helped both further progress and learning American English.

Language of science: one should distinguish spoken and written. The former is used for teaching and for communicating results to fellow-scientists in seminars and conferences. The latter is used in writing articles in scholarly journals and in books. Being brought up in the French language and culture allowed me to give primacy to the written word. I could only admire the much more verbal-oriented and talented American colleagues.

The former carried considerably more importance during my career, it makes up the record, the sum total of one's published work. As the Latin proverb has it: *verba volant, scripta manent* (words fly away, writings endure). I paid attention to what I wrote and tried to do it competently. Of course, there were examples for me to try and emulate, other chemists who were master stylists in English.

Pride of place goes to Robert B. Woodward (1917–1979). He paid the utmost attention to crafting not only impeccably, but memorably too. In Jeff Seeman's (b 1945) words, "This style is figurative, emblematic, and highly ornate with a heightened emotional tone evoking strong feelings in the readers." It could be deemed pretentious; yet RBW could get away with being grandiloquent since scientifically he towered higher than any other chemist in his time. I do not wish to add here to the hagiographic literature about RBW's eminence—he was indeed a great scientist and a very major chemist.

I wish to add a small note though, that wont take away any of his huge merits. RBW's prose should be considered in the context of his having been on the faculty at Harvard—and justifiably proud of it. Hence, one should mention other Harvard chemists who also wrote handsomely: James B. Conant (1893–1978), who mentored other distinguished Harvard chemists; the man-and-wife team of Mary (1932–1977) and Louis F. Fieser (1899–1977), with Mary holding their incisive pen; Frank H. Westheimer (1912–2007), who wrote an acclaimed report (1965) on the state of American chemistry; Paul D. Bartlett (1907–1997), who fathered an entire school of world-class physical organic chemists; and E. Bright Wilson (1908–1992), a luminary of physical chemistry, who published in 1952 a highly influential book, *An Introduction to Scientific Research.*

I cannot help mentioning another stylist of note, a Woodward contemporary, who deserves attention for the cogency of his writing, William von Eggers Doering (1917–2011). As he said in an oral interview, "I write with no style whatsoever, but I pay a tremendous attention to writing in such a way that what I'm trying to say can't be misunderstood."

Which does not rule out a measure of originality. For instance, the abstract of a 2002 article starts thus: "This paper addresses the decades-old problem of gaining a measure of intellectual control over the fate of the diradical intermediate in not-obviously-concerted thermal rearrangements."

The title of a 1999 article includes the meant to intrigue question "Chameleonic or Centauric Transition Region?"

Doering knew to whet the appetite of readers: consider this sentence, at the end of the abstract of an article in 1963: X "is a uniquely interesting molecule in which individual carbon atoms must circulate freely about the structure quite independently of each other if the Cope rearrangement operates."

The ugly face of war always lurks. I have just mentioned Louis Fieser, who with his wife Mary Fieser and her deft pen authored many a nice manual of organic chemistry. He had devised in 1942 a highly flammable and sticky gel named napalm. It was used during World War II and would be used again—thus deserving place in infamy. We fellow-chemists shuddered at the knowledge that Fieser was proud enough of his invention to collect articles in the press reporting uses of napalm.

Indeed World War II was still very much in our collective mind during the Sixties. At Princeton, some of my older colleagues, Walter J. Kauzmann (1916–2009), Charles P. Smyth (1895–1990) or John Turkevich (1907–1998) had worked on the Manhattan Project. And, in the context of the Cold War, some of my younger colleagues had close associations with the military. Most noteworthy was my contemporary John M. Deutch (b 1938)—he would go on to head the CIA in 1995–1996—who at the time targeted cities of the USSR for nuclear annihilation.

Now to recollections of my near-involvement in a number of military conflicts, which I'll set chronologically. First, there was the Algerian war. My age at the time made me vulnerable to be drafted. During a skiing holiday in Morzine, about 1956, I lined up a pass (col de Coux) through which I would escape into Switzerland, if I were indeed be called upon to fight in that war.

In later 1956, the insurrection against Soviet rule occurred in Budapest. By then, the parents of my second mother, Ella, had been able to leave Hungary and were safe with us in Grenoble. Nevertheless, she still had friends in Budapest, the Eleks and the Varkonyis, who escaped and found shelter in Canada. Repression was rather swift in crushing the revolt. And the new government headed by János Kádár (1912–1989) provided somewhat less harsh a Communist rule.

In 1958, the Fourth Republic ended and De Gaulle came back to head the country. The Left screamed bloody murder and branded De Gaulle a near-putschist. For a few weeks, things were tense.

In April 1961, some of the top French generals in charge in Algiers rebelled and staged a putsch against De Gaulle. Had it been successful, my having been born in Algiers—and my parents, for their own reasons, having not bothered to register their move in 1940 from Algiers to Grenoble—I might have been drafted into that war: by France against Algerian independence. But this insurrection fizzled out and De Gaulle continued to be firmly in charge.

My then wife Martine and I arrived in Princeton on Labor Day, September 3 1962. Thus, we lived in the US during the Cuban Missile Crisis and were actually on a sightseeing/shopping excursion to New York City on the most tense day, Saturday October 27th, termed the most dangerous day in the history of mankind.

As written about earlier in this book, Martine and I returned to Princeton in the fall of 1965. My employment by the University as a foreigner gave me immigrant status, I was eligible for the famous Green Card as a permanent resident. It also made me eligible to fight in Vietnam—the American consul in Paris informed me, very seriously, when I visited the Embassy to have my passport registered and stamped. This very congenial young man, quite young, then proceeded to tease me about going to Princeton—he was a Harvard man. And indeed, a few weeks after my arrival, I was notified I had to register with the Draft Board in Trenton—the administrative capital of New Jersey. Fortunately, I was not sent to Vietnam. It was not even a close call: Robert F. Goheen (1919–2008), who then led Princeton University, wrote a superb letter to the draft board, telling them I was more useful to the US in teaching American students.

Princeton in the Sixties was not exempt of the turmoil that swept through universities worldwide. I have already mentioned the quick and efficient reaction of Kurt Mislow, chairman of the chemistry department, about my report of May 1968 in Toulouse. And Princeton's attempt at increased diversity in its student body brought a few problems: on March 11 1969 African-American students occupied the New South administrative Building on the campus.

Twenty years later, in 1989, the USSR as a country ended. The Cold War, through which I had passed personally unscathed, was over—at least temporarily.

In closing this segment, I ask myself whether my work as a chemist was of any use to the military, in any country. I do not believe it to have been the case. I made a contribution—big or small, this is not for me to say—to the advancement of knowledge. Much of it had to do with catalysis of organic reactions by inorganic solids. My work (unpublished) with Keith Pannell on storage of heavy metals in clay matrices, if applied to radioactive wastes, contributed to peaceful uses of nuclear energy.

Now to an anecdote: a former coworker who has remained a close friend, went from my lab in Liège to a postdoctoral appointment elsewhere. He and several lab-mates contracted severe allergies. They launched their own inquiry. There was a burst of skin rashes and other manifestations whenever someone opened a certain refrigerator. They proceeded to pinpoint a chemical kept there in a flask. What to do? To notify the head of the lab was iffy, he might be sorely tempted, given the awesome activity of that substance, to tell the military. They did then what I hold to have been exemplary: they got rid of the stuff and all the entire documentation about its preparation.

And thus Lucky Pierre went through the twentieth century without having had to suffer directly from its numerous wars and appalling carnage.

As I started my education in chemistry, during the late 1950s, I was taught some glassblowing and how to pierce cork stoppers—technical knowledge that I never had to draw upon, chemistry having moved on quite a bit.

In the Sixties, a chemistry laboratory, whether in France (Gif-sur-Yvette) or the US (Princeton) showed a set of benches and a hood. Most of the work was done at the bench. A shelf above it carried bottles of organic solvents (such as acetone, hexane, carbon tetrachloride, ether, dimethyl sulfoxide), filter paper, small glassware (funnels, erlenmeyers,. . .).

The mere mention of such diverse items raises the question of the suppliers. There was a gamut, between the highly specialized and the more general. In addition, some entities such as the chemistry departments I worked at in the US featured stockrooms. In France, Prolabo in Paris—an offshoot of the Rhône-Poulenc company—supplied laboratories.

To venture a generalization, during the 1960s–1990s, most chemicals came from American companies, Aldrich and Eastman foremost. Much of the big instrumentation—nmr and mass spectrometers, infrared spectrometers, gas and liquid chromatography apparatus—was provided by American companies: Varian, Perkin-Elmer, Waters, . . . In addition, there was a more specialized niche for laboratory suppliers from small countries in the Western world. Two examples: constant temperature oil baths from the Tamson company in the Netherlands, Mettler precision balances from Switzerland.

Swiss manufacturers specialized indeed—no surprise in the country renowned for its watchmakers—in precision instruments. Another example of ubiquitous Swiss-made presence in laboratories during the Sixties and later to some extent, the Dreiding stainless steel molecular models. Those had been designed by André S. Dreiding (1919–2013), a professor of organic chemistry at the University of Zürich. I vividly recall walking into E.J. Corey's (b 1928) office at Harvard, he had above his desk, within easy reach on a kind of clothesline, an array of already assembled Dreiding models for the molecules he was then (1966?) working on, their synthesis as his goal.

I knew Dreiding personally, we met on a number of occasions, first and foremost at the Bürgenstock conferences on stereochemistry. They were his idea originally and his colleagues across town at ETH were quick to embrace the idea and more or less capture it for their benefit. André and I still exchanged end-of-the-year greetings in 2012, he was to die a few months later.

To convey how organic chemistry during the second half of the last century was a small world—and how delightful that was—I'll come back to Aldrich, American supplier of chemicals. Alfred R. Bader (1924–2018), a former student of Louis Fieser, had started that company in 1951. It had expanded considerably and its catalog offered tens of thousands of different chemicals. My guess is that Aldrich was the number 1 supplier of academic laboratories—probably not only in chemistry. I met Alfred Bader for the first time during one of my stays in Ithaca, when he came to give a lecture: Bader was not only a chemist and a businessman, he was also an outstanding amateur art historian. He and his wife Isabel owned an art gallery in Milwaukee. We befriended one another and exchanged occasional correspondence—he did not fail, every single time to present me with an offer to purchase some artwork from his collection.

Was I an acceptable supervisor of scientific research? At least, I gave it my very best. Early on, I received sound advice from Jay K. Kochi (1927–2008), then at the University of Indiana. He told me of the multiple roles a group leader needs to play, half-a-dozen at least, guide, boss, banker & money lender, counselor, confessor, psychoanalyst, . . . What I found out on my own was the ease—not the complexity—of the task, given that one is dealing with people: each is different and provided you respect the differences from yourself and have genuine empathy, you can achieve the ideal of maieutics: most graduate students and postdocs gave me their very best.

These in my opinion are the qualities of a supervisor of science. Patience—trust—mutual respect—attentiveness. Make yourself available at all times. In short, bring to this part of your job the same qualities and character demanded in a successful marriage. Treat each of your coworkers as if he or she were a life partner.

Need I add that, by and large, I skipped the laboratory notebook formality—even though it was considered in the profession as an unavoidable and a sacred duty?

But I have to be more specific and offer some capsule portraits; they will point some flaws in coworkers who otherwise were admirable.

A was a reserved and shy person. He worked methodically and hard. He saw his task as filling holes in the literature which he (mistakenly) trusted 100%. Hence, each paper of his—at least once he had achieved his independence from my supervision—depended on existing work from other labs, a subjugation that was never a temptation of mine. A healthy dose of skepticism is needed in perusing the literature.

B was highly idiosyncratic. He was both brilliant and fiercely independent. Somehow, he was able to pursue only his own suggestions and lines of inquiry. I found out that the lightest of touches was needed with him, even though he paid respectful lip-service to my indications and directions. To put it another way, we achieved at all times a truly creative dialog.

C was a born wit and a tease, he loved to needle me. We became good friends. He was admirable in entering unknown areas, which he loved to venture into and had the courage and intellectual ability to do. He could read and he could write difficult papers, for which he had the patience and the didactic ability.

D was neither very bright nor intellectually brave. But he was hard-working and he had admirable working habits. We complemented one another well, he trusted my intuition. He was steady in the lines of work he espoused and that became very fruitful as a result.

E was a character, a unique character. For years, I despaired of having him complete his doctoral work. He had, then, the infuriating habit of letting his chemical samples age—out of curiosity. Of course, the ongoing processes of oxidation, polymerization—rotting in short—got in the way of what was meant to be studied. What he needed—I realized after quite a while—was a problem of first importance. Then he became absolutely brilliant.

F was also a character. Arguably the smartest person I ever came across. At a very young age—obviously much too young—he was tossed from an Indian Brahmin family into arduous graduate study in the Ivy League. He chose instead to spend his entire time reading historical novels by Alexandre Dumas that I introduced him to. He devoured them. This kind and most congenial man became a very successful trader on Wall Street. He is now back in a chemical laboratory, to fulfill his vocation at long last.

G was a technician. He thus lacked the degrees to be allowed into undergraduate study, let alone graduate study. But he was enthusiastic and eager to join. He was treated as an equal by my other coworkers. Their teamwork, in a significant number of studies, led to a number of publications in which he was, most justifiably, a co-author—incidentally, I have always listed the names of my coworkers and mine in alphabetical order.

H was an undergraduate. He was supposed to prepare a small original piece of experimental work during his senior year. But he was totally immature, a child who spent his time playing in the hallway at various games of his own devising. I refrained cracking the whip, and left him to his own devices. After graduation and acquisition of a BS, he attended a Bible School in the American South. He is now a professor of chemistry in a major university, most eminent in his chosen field of science.

J came to my group having already a degree in applied science. He loved manual work. He was excellent also at experimental work in chemistry. After joining my research group and getting his Ph. D. from Princeton, he followed me to the University of Liège and he helped me set-up my lab there. He went on to postdoc with a future Nobel prizewinner in chemistry. However, a career in chemistry, academic or industrial, did not carry enough appeal for him. After thinking it over, he radically changed course and enrolled in the medical school in Liège—for several additional years of study. He enjoyed a successful and rewarding career as a dentist.

K was the dream of any group leader, a lieutenant gifted for management, with its multidimensional tasks. Accordingly, after his well-earned Ph. D. under my supervision, he became a university president and kept at it most successfully for decades.

On reflection, these stories and others which I could also have told demonstrate all the good that can ensue from letting people under one's supervision enjoy total freedom, have a free rein to run, and run their lives as they please.

On to the key topic of peer groups. How to define them? What kind of a peer group did I belong to? How did it influence my career? These are some of the relevant questions.

My definition is: the group of colleagues you know personally, more or less on a first name basis and that you meet several times a year at conferences. The group I belonged to was made of organic chemists, mostly older than I, from the US and Western Europe. It numbered 60–80 persons, 10–15% of whom were women—in my time of 1960–2000.

A peer group is foremost your audience and readership, the people who attend your lectures and read your papers—they are also called upon to referee them and vet them for publication. I wish I could go through a list of their names—many of them became good friends—but this would detract from the purpose of this chapter, reminiscing in general and not in particular. I shall put it this way, in yet another definition of a peer group: the sum total of colleagues one not only admires, but also carries a warm mental image. I shall submit something of a more operational definition: a peer group consists of the people one meets at small conferences, those with small attendance, a restricted roster of, say, several dozen participants at most. Hence, this segment will focus on very few of the persons that made up the group I belonged to, on those that were important, either to my development as a scientist, to my academic career, or both.

Jacques Reisse (b 1936) from the Free University in Brussels has been a close personal friend throughout my career. I owe him my call to a professorship at the University of Liège in 1970. While both in Belgium, we met often for seminars and to attend lectures by leaders in our joint profession. We would also meet at annual conferences, especially those held at the Bürgenstock, in Switzerland. He did outstanding work and his written output is most meritorious.

Guy Ourisson (1926–2006) was a guide throughout my career. We had met at the end of 1963, when he invited me to the University of Strasbourg for a most lively lecture on "Should one publish in English?" He was impressed by my *Leçons de chimie* books, enough to intervene for my appointment (1986) at *École polytechnique*. In the meanwhile, we met often, often at yearly conferences such as the GECO (*Groupe d'étude de chimie organique*) he had created in France on the model of the American Gordon Conferences and the Bürgenstock Conferences on Stereochemistry. I attended and lectured in at least half-a-dozen in each of these series.

Another, slightly older than myself, remarkable scientist and close friend who acted as a mentor and a tutor was Pierre Potier (1934–2006). I have told earlier on of his devising two widely used antitumoral drugs, Taxotère and Navelbine. He was an example of intellectual freedom, with his disregard for Orwellian Newspeak and conventional behavior.

I owe him attendance of several meetings of the French *Groupe d'étude structure-activité* (GESA)—patterned after Guy Ourisson's GECO. These GESA conferences were singled out for me by an amazing figure, a very great scientist and thinker, a charismatic figure too, Henri Laborit (1914–1995). He amply deserved a Nobel prize in medicine. We owe him; in surgery, potentiated anesthesia, lowering the metabolism and body temperature (so-called artificial hibernation). He was responsible for introducing chlorpromazine as a pioneering antipsychotic drug. At the time of the GESA meetings that I attended in which he was present, his main interest had become neurochemistry of the brain—he convinced me that, if I had to resume a chemical career, I should enter that subfield. This was the time when the neurotransmitter gamma aminobutyric acid (GABA) became of prime importance. Current studies of the brain, those using in particular functional MRI (magnetic resonance imaging), are in the direct lineage of Laborit's inspiring ideas.

An attraction of conferences, not only to myself, was the presence of flamboyant lecturers. One such person, already mentioned, was Henri Laborit. His memorable presence at the GESA meetings, his interventions during discussions in particular, made them worthwhile for me.

Let me try to summarize his view of the human brain. Taking his cue from cybernetics, Laborit described the central nervous system in terms of levels of organization, from the molecular and cellular to the cerebral, psychological and social. Their mutual interaction is subject to feedback loops, through which an organism and its parts maintain their structural integrity. Laborit brought attention to cortical parts of the human brain that distinguish it from earlier forms in evolution. Nowadays, important studies, such as by Stanislas Dehaene (b 1965) and his team, use magnetic resonance imaging (MRI) to visualize the brain at work. Thus, brain regions that connect with one another in specific tasks—such as reading words or verbs generation—can be identified. Then and now? All this current research activity can be traced back to Laborit, his influential teaching and his books of popularization.

He was indeed flamboyant at meetings. This adjective's etymology points to its meaning, i.e., literally emitting and inducing flames in one's listeners minds. Indeed, to evoke flamboyant colleagues in chemistry I can do no better than to name artistic performers showing similar characteristics such as, among classical musicians I listened to, the pianists Glenn Gould (1932–1982), Sviatoslav Richter (1915–1997), Alfred Brendel (b 1931), violinists David Oistrakh (1908–1974) and Igor Oistrakh (b 1931), cellists Pau Casals (1876–1973) and Mstislav Rostropovich (1927–2007), flutist Jean-Pierre Rampal (1922–2000), etc.

Whom then among my direct colleagues in chemistry did I find flamboyant in presenting their work? I'll mention very few only. Ron Breslow (1931–2017), from Columbia, to whom we owe biomimetic chemistry; and who was denied a Nobel prize because of fraud by a coworker. Harry B. Gray (b 1935) at Caltech. And Bill Lipscomb (1919–2011) from Harvard.

Lipscomb lectured at the Bürgenstock conference in stereochemistry that I attended, in 1970. His presentation was fine and even elegant. What was electrifying though was his presence throughout the meeting: whatever the topic of the other lectures, during the following discussions, Lipscomb made comments that were, as a rule, both competent and incisive. He was a chemist with a sound command of the whole discipline! The catholicity of his interests was admirable. At the time, he was clearly running for a Nobel award, which he deservedly received not long afterwards, in 1976.

That was then. What about now? Flamboyance has been replaced by hype; in other words, advertising as a marketing ploy has replaced communication of one's results to peers.

Was I lucky to have known such a great mind as Duilio Arigoni (1928–2020). This Swiss chemist spent his entire career at the Swiss Polytechnic (ETH) in Zürich. He was from Ticino, in Southern Switzerland. He once confided to me that as a student and even young faculty member he was taunted by the Swiss Germans for being a "macaroni." Such insulting denigration may have had something to do with his overall excellence in ths public eye, whether from the written or the spoken word.

My first images of Duilio were when he came to lecture in Gif-sur-Yvette, at the *Institut de chimie des substances naturelles*, where I was preparing my doctorate in the early 60s. The rumor that greeted him was that this still very young scientist would for sure win a Nobel prize.

Over the years, I met him many times, often at conferences, in France, Switzerland and the US. I'll select two such encounters from my album of memories.

Memory number 1 is from Princeton. I had been invited for dinner at the Mislows, Kurt (1923–2017) and Jacqueline (b 1935). Kurt viewed me as not only a young colleague in the department he was then chairing, somehow my culture shone in his eyes. I knew it because of an earlier dinner invitation, in April 1966, in his home: also a small party, with six of us, Kurt and Jacqueline Mislow, Martine and me, the other two being the German writer Hans Magnus Enzensberger (b 1929) and his young wife. They were fresh from a trip to Castro's Cuba. Enzensberger had come to Princeton to attend the meeting (it made history) of the Gruppe 47. It showed Kurt's trust that I was capable to sustain a multi-sided conversation.

A year or two later, Duilio came to Princeton, as part of a lecture tour. Again, I was a dinner guest of the Mislows. I was by myself, Martine was on duty that evening at the hospital. After a while, Jacqueline busied herself in the kitchen. The three of us conversed, on many a topic. What I thus witnessed was a rivalry of wits between Duilio and Kurt. Obviously, they knew one another all too well. After his Ph. D. at Caltech with Linus Pauling (1901–1994), Kurt had gone to Zürich for a post-doctoral stay with another (future) Nobel laureate, Vlado Prelog (1906–1998). There he had come to know and befriend Duilio. Kurt was a deep thinker, someone who has enlarged the perimeter of chemistry. But what I witnessed on that evening, until very late, way past midnight, was Duilio running circles around beleaguered Kurt. He clearly was much smarter. He had a scalpel-like intellect. His was a beautiful mind, to borrow from the title of Nash's biography.

As an aside, since I have just alluded to the Nobel prize regarding both Pauling and Prelog, why is it that Arigoni did not receive one—which he amply and superbly deserved? I do not have an answer. What I can volunteer, again as a witness, is a rather indirect complaint by Duilio for not having been thus recognized. In a conversation, some time in the Seventies, he exclaimed rather bitterly that the Swedish Academy again had denied Jorge Luis Borges (1899–1986) the literature prize he so obviously deserved (I concurred).

Memory number 2 is from a GECO meeting held in an hotel in Saint-Gervais, Haute Savoie. Robert Woodward was in attendance. Duilio had also come and brought his wife Carla and their children. It was during a summer, possibly in early September. Weather was beautiful and warm. In the end of the afternoon, we participants were gathered on the lawn, around the swimming pool. The organizers brought out a portable blackboard and set it up. There were a few chairs, the rest of us sat down on the grass. And a session of brief communications—*impromptus*, in principle improvised on the spot, started. One of them, by Jean-Claude Jacquesy had to do with steroidal chemistry. It was quite good and interesting. As soon as Jacquesy was done, Duilio jumped to his feet and said that he had done related work and wished to tell us about it. He was amazing: for 15–20 min, he reeled off from memory reams of precise data, and made a memorable presentation—positively brilliant.

One of my delights from academic life came from witty colleagues. Legendary among us for his stories and jokes was Vlado Prelog—many of us regretted that he refrained from such gems in his autobiography. He was influential among other colleagues at the ETH in Zürich. Duilio Arigoni was one of them. He too was delightful for the jokes he told. I recall two such Swiss jokes.

Two Swiss gentlemen are in a Parisian restaurant, ordering their meal. Which wine to accompany it with? They scan the wine list for Swiss Dole and are unable to find it. Their waiter asks them for their wine choice. They answer Dole. The waiter, knowing that they do not carry it, goes to the sommelier, who tells him, "nothing simpler: get a decanter, put in some mustard, salt and pepper, vinegar and fill it with water. Bring it to them, this will work well." The waiter dismayed nevertheless does as told. The Swiss men sample this mixture. One says to the other, "I knew it, they save the best for export."

And here is the second joke that I heard from Duilio and that I recall, even after many years: God has nothing to do and is a little bored. He decides to descend to Earth, for a small inspection. He thus finds himself in the mountain country of Switzerland. The weather is sunny, there are cows in the pastures and vineyards in the distance. An old peasant stands nearby. God approaches him and starts a conversation, "How are you, how is everything?" The old man answers, "not bad, things could be worse. My wife and I are getting on in age, but we enjoy good health. I have little to complain about. See how it is beautiful here? I own a few cows, we have also some vines, chicken and a vegetable garden. We are not wealthy but we lead a good life". God looks around, everything indeed is beautiful. He feels very happy with his creation. The old peasant then says, "would you like to sample our milk?" God answers in the affirmative. The old man serves God a glass of milk. He sips it slowly in the sun, his eyes feasting on the lovely sights. He feels wonderful, unworried and, yes, very much satisfied with everything he has created. But he ought to proceed with his inspection tour. As he prepares to leave, he tells the peasant how enjoyable this little stop has been. And the peasant answers Him: "it will be 2 francs for the milk, Sir."

Prelog and Arigoni both knew how to spin a yarn: these born raconteurs knew how to keep talking, making up descriptive details as they went along, apt at sustaining interest, egging-on their listeners until they dropped the punchline.

We chemists also loved, when we got together, exchanging Norbert Wiener (1894–1964) and Paul Dirac (1907–1984) stories. The former was legendary at MIT for his absent-mindedness. Such as when he arrived in a classroom for teaching a lecture, pushed aside someone in front of the blackboard and started lecturing to the people in the room. He was in the wrong classroom.

From Dirac, who also was in a classroom having just finished giving a lecture; a person in the audience stood and said that he had not understood a particular equation. Dead silence. No answer. After a while, the chairperson told Dirac, "would you care to answer that question?" And Dirac said, "it was a statement, not a question."

And then there were the elder statesmen of science. Was I lucky to have known some of them and interacted with them.

Whom do I recall? Frank H. Westheimer (1912–2007). He was a true gentleman and a scholar, the embodiment of that stereotype. In addition, he opened a novel subdiscipline, in-between organic chemistry and biochemistry. If nowadays we witness a merger of chemistry and biology, it is to some extent his legacy.

Fred Bordwell (1916–2002). To me, his sight is undistinguishable from his urbane demeanor. He always wore a suit and a tie, and his full head of white hair had him nicknamed "the silver fox." He was mild-mannered, he was dignified and he was extremely knowledgeable.

Harold Hart (1922–2019). He was such a gentleman and an admirable colleague. When I visited Michigan State University in April 1981 and gave a seminar in chemistry, he was my host: urbane, congenial and warm, a delight to be with. I recall, in particular, his showing me the mural that Charles Pollock (1902–1986) had painted during the Thirties on a nearby building. He did outstanding chemistry and was most modest about his highly significant achievements.

Marshall Gates (1915–2003), in addition to being an outstanding organic chemist, the first to synthesize morphine, was an superb editor of *JACS*, the *Journal of the American Chemical Society*, fair to everyone and uncompromising as to the quality of the contributions. I met him by occasion of a lecture at the University of Rochester, he was unassuming and very modest. I learned from his obituary of "his athletic pursuits as a sailor and skier. He kept at the craft of glassblowing, making many of his instruments for laboratory use, built both model and small wooden sailboats, and enjoyed a weekly poker game for more than 35 years."

Gilbert Stork (1921–2017) hosted one of my visits and lectures at Columbia. Born in Brussels, he exuded Old World charm, was urbane and witty—in addition to a splendid track record in chemistry, to which he still added significantly in his old age.

Paul D. Bartlett (1907–1997), from Harvard like Frank Westheimer, was scientific father to a whole crew of world-famous physical organic chemists. At the time the classical-nonclassical ion controversy was raging, he published a book, *Nonclassical ions*—actually a collection of wisely and sharply commented papers—that I have consulted very many times in preparing for a class or writing an article.

Wallace S. Brey (b 1922) was extremely modest and never tried to impress, even though he had been among the pioneers of nuclear magnetic resonance in chemistry. He was soft-spoken courteous, the true Southern gentleman. He was the founding editor of the *Journal of Magnetic Resonance*: he steered it admirably, from the start it was a most distinguished journal of high caliber.

I was lucky to have met and talked with Luitzen J. Oosterhoff (1907–1974) when we both attended in the late Sixties several of the Bürgenstock conferences in stereochemistry. He was a Grand Old Man of quantum chemistry, which he had taught and practiced at the University of Leiden. He had been a precursor of the Woodward-Hoffmann rules for electrocyclic reactions, without realizing how important these would turn out to be. I was much impressed with his continued enthusiasm for chemistry and his recollections of earlier chemists who had faded into oblivion in spite of noteworthy contributions.

To close this segment, I turn to a couple of French chemists who have also markedly impressed me with their outlook in old age. Emmanuel Grison (1919–2015): always impeccably dressed, his physique remained that of a young man, flexible and quick in his movements. He unfailingly showed great deference to anyone, regardless of status. He was indeed a man of exquisite urbanity, always courteous, always benevolent, and of a lively intellect welded to a vast culture. We shared an active interest in the history of chemistry. Thus, to give an example of his erudition, when I told him about the book by Octave de Ségur, which reflects the chemistry teaching provided at *École polytechnique* at the beginning of the First Empire, where the author uses a rather peculiar chemical symbolism, M. Grison immediately identified it as Hassenfratz's (1755–1827), about whose life and career he was the recognizedworld expert.

He later asked me to speak at a symposium at the Chemistry History Club, which he had helped found. In addition to his biography of Hassenfratz, historians owe him an edition of the correspondence Kirwan-Guyton de Morveau, of which he was one of the co-editors.

To close this miniportrait, an anecdote testifying to his generosity, even more than to his rapidity of thought. In a conversation at lunch, the discussion turned to lecturers to be invited to the École. I mentioned France Quéré (1936–1995), she died way too early, the great Protestant theologian, also the wife of Yves Quéré (b 1931). He exclaimed, "Definitely not. she is already too much in demand. She would feel obligated to agree. You can't do that to her!"

In short, he was a righteous person, whom I am honored to have met and of whom I retain an admiring memory.

To close this section devoted to exceptional persons who brought together the stature of a great scientist and the highest morality, a word on Alain Horeau (1909–1992), whom I was honored to have known. I wont belabor his accomplishments, they have been documented by others: the method he devised for determining the configuration of an optically active molecule; his holding the top position of administrator at Collège de France; presiding the French Academy of Sciences; and fathering 12 children.

As a professor at Collège de France, he informed the community of Parisian organic chemists of the important advances in the field that were being made worldwide. His documentation was exemplary. A small crowd showed up on Saturday mornings to attend his lectures. Thus, together with Edgar Lederer (1908–1988) in Gif-sur-Yvette who during the same period organized lectures in Gif by leaders in the field, in a time of blight they saved the honor of French chemistry.

I recall also a decisive move of his. Someone, whom I shall leave nameless, was elected to the presidency of a professional society by stuffing the ballot boxes, Horeau stepped in, made this person resign and organized new and fair elections.

It did not escape my attention that the gentlemen I have just featured—no ladies, I regretfully add, due mostly to the mentality of the times, pressuring them into subservience—on the whole lived into their 90s. It would be a gift from the Gods were I to follow their example.

Now and Later

25

I enter here well-trodden and yet forbidden territory: statements to the gist of yesteryear having been lovely and fantastic in contrast to the present time being abhorrent. I shall strive not to belabor the point, obvious to me unfortunately. Instead, I shall explain my reasoning. More importantly, and here I am addressing future historians, I shall attempt to characterize the *episteme*—the *Zeitgeist* if you prefer—I was part of during the second half of the twentieth century.

We scientists believed in science. It was not a naïve belief, we knew it could not cure the ills of the world. To express it succinctly, we deemed important the advancement of knowledge. Why? Because, together with art and literature, it boosts the human spirit. Yes, we cared more, much more, about self-improvement than about a quixotic attempt at improving human society.

To which, in complement, came the exhilaration of being part of the closely-knit worldwide scientific community. In the previous chapter, I mentioned some of the most appealing traits of fellow-scientists: intellectually playful; scalpel-sharp analytical minds; beauty lovers and admirers; master stylists of English prose; and, last but not least, moral elegance.

What then were some of the factors for the spoilage that gradually befell chemical science, starting in the Seventies and Eighties? The first I'll mention is due to the dual nature of chemistry, both a science and an industry. We academic chemists had repeatedly to try and seduce back a public that was repelled by the disasters that our industrial colleagues had either wrought or been unable to prevent. To mention a few: the ozone hole, due to chlorofluorocarbons; global warming, due not only to carbon dioxide from the burning of fossil fuels, but also to methane (natural gas) escaping into the atmosphere; DDT and pesticides in agricultural use; toxic food additives—a long list, eg bisphenol A, phtalates, perfluoroalkyls, perchlorate, synthetic dyes, nitrates and nitrites. Because of its diffused guilt about environmental pollution, automobile overuse, deleterious personal habits (alcohol, drugs, cigarettes) the public projected it onto chemistry and chemists. Plus a litany of catastrophes at industrial plants: Texas City in the US in 1947; Seveso, in Northern Italy in 1976; Bhopal in India, in 1984; the explosion at the AZF plant in Toulouse,

France, in 2001; ... Disasters which, needless to say, are complemented in the public eye by accidents at nuclear plants: Three Mile Island (1979), Tchernobyl (1986) and Fukushima (2011).

These terrible events fed chemophobia on the part of the public. It had other causes too. One was the internal language of the chemists, yes the language of formulas, that they prided themselves upon—all too smugly. Let me offer this example: the widespread practice, in the classroom, of the paper tool of reaction mechanisms, taught with Lewis structural formulas, using curved arrows to denote motions of electrons. This practice, while communicating the rational understanding of chemical reactions and their underlying logic, can also be seen by some students as a modern counterpart to medieval scholastics. To put it another way, it had some of the features of slang with respect to more thoughtful and dignified speech. And it may have bred cynicism and skepticism on the part of students who see this paper tool turned into a universal explanatory device.

That chemists worldwide can communicate with one another using graphic formulas brings attention to them sharing what can only be termed a tribal culture. Speaking now as a full-fledged member of the tribe, we interpret the experiments we devise in terms of molecules and their behavior, i.e., by reference to invisible entities. To us chemists, such ghosts not only have a presence, we enter into daily interaction with them. We see them in our mind's eye.

Chemists learn their craft from serving a long apprenticeship, in like manner to a gilder or a professional musician. They usually undergo slow ripening into an alert and experienced professional. Which explains why chemists are older than, for instance physicists, when earning a Nobel prize. Such late blossoming was exemplified in the Seventies by the elegant biomimetic syntheses achieved by William S. Johnson (1913–1995), then in his mid-to-late 60s, on the verge of retirement. He reminds me of Georg Friedrich Haendel (1685–1759) who composed "Messiah" when he was 56—an old man by the standards of his times. The community of chemists rejoiced in Johnson's achievements, he deserved praise for his burst of creativity at his age. There had been likewise affectionate respect for Joel H. Hildebrand (1881–1983), who taught at Berkeley, and aged 100 continued to be active as a chemist.

This tribal culture of chemists has other attributes than a language of their own and the slow ripening. It is characterized also by the somewhat suspect rooting of the science in alchemy—even with the passing of centuries, it is an inheritance evoking witches and the scaffold. Chemical laboratories appear to the public as all-too-prone to dangerous fires and explosions. Not to mention strange if not mephitic smells. All such attributes also feed chemophobia. An apt comparison is to free-masons, whose rituals and secrecy feed suspicion.

In addition to chemophobia, more generally an anti-science mentality developed towards the end of the twentieth century. It had deep roots, such as resentment towards neoliberal economics, American dominance and, probably most interesting, the return of religion that André Malraux (1901–1976) had forecast as an attribute of the twenty-first century.

Sociology of science made headway during the second half of the twentieth century. To the extent, worth plaudits, that social studies of science have gained legitimacy among the humanities and social studies in British universities. Some sociologists of science were swept by the anti-science mood. Accordingly, they have resurrected a neologism, technoscience, and routinely used it as a term of abuse; an amalgam reminiscent of totalitarian propaganda.

Nowadays, no day goes by without my receiving a request or several to contribute an article to a journal I had never heard of, or present a communication at a conference to be held, generally speaking, in a Pacific rim country. I wonder what participants at such 'scientific' conferences look like; a bunch of octogenarians in wheelchairs cheered by a larger group of Third World delegates, maybe?

The invitations I get are clearly aimed at my authorial vanity. Were I to accept, is the underlying message, my name would draw in other colleagues. Of course, I do not even bother answering and consign these emails to the spam heap.

This is an offshoot of so-called predatory publishing. Its appearance, its inordinate swelling into a tide, postdates my retirement in 1999. The roots, though—the rot one might say—go back much earlier, to the Sixties and Seventies. Some regular commercial publishers in those remote times—Pergamon Press for instance—were in the business not only to provide scientists with a needed service, but also to make money hand over fist.

Predatory publishing, as its name indicates, preys upon legitimate scientific publishing. The latter being a $10 billion industry whose leaders enjoy double-digit profit margins, the former can be guesstimated as a $ 1 billion industry with commensurate profit margins.

Legitimate scientific publishing preys upon academic libraries by selling them subscriptions to scholarly journals. Libraries are committed to their purchase in order to provide research scientists with the documentation they need—not to mention the satisfaction of seeing themselves in print. Thus, there were no fewer than about 700 journals in chemistry in 2018. In 2019, the average cost of a chemistry journal to a library was about $ 6000. Hence, a library aiming at holding only 30% of the journals in chemistry has to budget of the order of about 1.2 million dollars. No wonder predatory publishers eye eagerly such a bonanza.

Of late, there have been paper mills too: churning out fraudulent scientific papers. The Royal Society of Chemistry recently (beginning of 2021) withdrew 70 such articles.

But the landscape I have just sketched out is already outdated, because of the Internet. The World Wide Web has changed the rules of communication, including scientific communication. To suggest what the dollar (or euro) value of our instant access to results from scientific research may be, I have done a small test. I went to PubMed, the bibliographic tool of the American National Institutes of Health. I typed in "cavernous angioma." The result, as of April 2021, is the existence of more than 9000 publications on this topic. Of this number, 1622 i.e., only 18% are freely accessible via PubMed. How much does it cost to obtain a reprint—in pdf format typically—of the others? € 50.00 for a typical article published in 2013, $ 36 for a 2015 article and also $ 35 for a 2019 article—these three making up my sample. As a scientist myself, I would guess that any half-way decent bibliography on this topic should encompass at least 50 papers. If these indeed show the same proportion of free and for purchase reprints, its cost is $ 41 x 35 = $ 1435—charged to the grant subsidizing the work. Assume there are a thousand health professionals worldwide working on cavernous angiomas, a not infrequent condition, and that niche alone is worth a million dollars in sales of reprints. Again, it is no wonder if predatory publishers eye with envy such a pot of gold.

And what about fake scientific conferences such as I get so often invited to help set-up? Take a look at the cost of a legitimate scientific conference. In 2019, June 20-24, the American Society of Microbiologists held its annual meeting in San Francisco. The Federal Drug Agency (FDA) paid for the costs of 32 participants at the level of $ 5900 per person—between registration, travel fares, hotel room and meals. Again, it is no wonder if such sums induce envy on the part of people and organizations intent upon making a quick buck.

At the turn of the century, has worldwide dominance of chemical science started switching from the US to China? A related question is whether demographics of scientists, by which I mean their country of birth, has changed markedly: are we about to witness a sea-change?

The answer to both questions is a flat no. Yet, from the onslaught of invitations by predatory journals and fake conferences, one might get the impression that China and Third World countries are already occupying central stage. The evidence to the contrary comes from—a Chinese source: a mere look at the Shanghaï ranking of the world's universities shows an amazing stability in time.

In 2020, the institutions on top in this rather authoritative ranking were, in sequence: Harvard—Stanford—Cambridge UK—Berkeley—Princeton—Columbia—Caltech—Oxford—Chicago—Yale—Cornell—UCLA. Had one fallen asleep in 1970 to wake-up half-a-century later, this ranking with 10 American universities in the world-leading dozen would have come as no surprise whatsoever—let's go back to sleep.

I am leaving aside the question of the factors explaining such a pronounced dominance by the American universities and Oxbridge. Another question then comes up: if China has not caught up yet, in spite of the will of its political leaders to do so and wrest number 1 status from the US, has it managed to do so more obliquely by nesting its progeny into the ranks of these leading institutions, as faculty members?

To gain at least the beginning of an answer I gave a close look at the composition of the faculty in the department of chemistry at Caltech. As an incidental, Caltech, number 7 in the 2020 Shanghaï ranking, was discriminated against due to its small size (number of undergraduate students) as compared to the other universities at the top. Were one able to correct for it, Caltech might well come up as number 1.

Anyway, it has currently (April 2021) 46 faculty people in chemistry. Of those, seven bear a Chinese name. Of those seven, only four were born in China and received their initial higher degrees (BS) there. The other three are Sino-Americans (as is well known, Asian-American families put a premium to make their children strive for excellence in education). What even a cursory look at the CVs of these

Fig. 25.1 My youngest brother Yves Laszlo (b 1964). An academic and a scientist too, he is a distinguished mathematician. In recent years, he was appointed director of research and teaching at *École normale supérieure* (2012–2019) and at *École polytechnique* (2019–present). He strives in these positions to raise the world ranking of these leading French institutions

seven faculty members at Caltech shows is that their earlier years as graduate students, postdocs and faculty members were spent in other elite universities among the dozen above-cited. Their credentials did not include Chinese institutions—which is not to say that, 20 years from now say, this situation won't have changed (but I doubt it).

Cambridge, Oxford and leading American universities: what explains such dominance in world ranking? Some features of the English language? Cultural factors? The latter, rather. Let me try to explain this superiority, durable as already stated (Fig. 25.1).

There is the money factor, obviously. Because philanthropy is an ingrained trait in the US, American universities are able to draw on sizeable endowments. Currently, Harvard's is 41 billion dollars. To give an idea of what this sum represents, it is three times the national budget of a country such as Tunisia! Which allow American universities to invest in both talented teachers and on the best infrastructure: classrooms, to hold classes for 25 students or so—as contrasted with the large lecture halls, meant for hundreds of students I have known in France, as student and teacher in succession.

Cultural factors are predominant, though. To pursue a comparison between French and American institutions, that I know first-hand, the former can be likened to a *jardin à la française* (a garden in the French style): their ideal is formatting students in a uniform way; whereas the latter aim for self-expression. The former depend significantly on rote learning. The latter on originality of thought. Gardens in the French style indeed connect with academia: I recall the advice I received from Lionel Salem (b 1937) when I came back to France as a professor at the *École polytechnique*: "watch out, here they cut anything that stands out."

Language comes in, too. I can state the difference thus: where French universities are keen on writing and textuality, their American counterparts conversely emphasize orality, in the form of discussion and debate. Let me repeat two points I have already made, Americans with their informality and congeniality are much more at ease with speech than the more formal French; and science thrives on challenging each new finding or idea, controversy is its live blood—this was cogently expressed recently by Dominique Raynaud (b 1961), in my opinion the best current French sociologist and historian of science.

This is not to say that Anglo-Saxon higher education does not put any emphasis upon the written word. It does. But the very form of the written essay differentiates it from its European—French say—counterpart. The American student is encouraged to provide a logically organized list of points (the so-called bullets in a Powerpoint presentation). French students are trained into writing three-part essays, stating in succession thesis, antithesis and synthesis.

While on the topic of the written word, what about the superiority of Anglo-Saxon publishing? Is it due only to the status of English as lingua franca, the present day counterpart to the hegemony of Latin, at the time of the Renaissance? Another relevant factor, I submit, is respect for originality and self-expression. It comes through in lovely little writing manuals, such as Strunk's and White's; also in copy-editing, a practice I became used to with American publishers, which I was dismayed to find their French counterparts lacked.

Yet another factor in the Anglo-Saxon superiority in higher education can be traced to the Latin saying or dictum, *mens sana in corpore sano* (a healthy mind in a healthy body). Sports are not only provided for in American colleges, there is strong pressure for every student to join and thus combine hard play and hard study.

The contrast in the legal systems reflects that in the educational systems. Anglo-Saxons rely on common law, i.e., the notion of precedent: each case is different and judges look to earlier cases that have set jurisprudence. The French judicial system is based on law, a set of rules deemed to have universal applicability.

Such difference between particulars and universals can be illustrated with a contrast proposed by Lawrence W. Wylie (1909–1996) in a lecture, I attended in the Sixties and still vividly remember.

A French child is first taught in primary school about *les points cardinaux*, i.e. the directions in space, North, South, West and East. Then comes the solar system, the relation of Earth to the Sun and to the other planets; the oceans and the continents; the various countries. With some luck, by the end of the year, this pupil knows where he or she stands in the universe at large.

An American kid is first told to draw a picture of his or her desk and of its position with respect to others in the classroom. Then, to give consideration gradually to wider and wider spheres. Until, ideally by the end of the year, he or she knows about his or her position in the world.

In other words, again, universals versus particulars. There is little doubt that, in a child's mind, the latter are easier to grasp and cope with.

The hegemonic dominance of English in scientific publications and that of Anglo-Saxon institutions of higher learning has obvious consequences on our lives. The impact upon economies has no need for belaboring, no week goes by without journalists mentioning American leadership. American influence affects everyday life, as well, from fast food joints to the position of GAFA (Google, Apple, Facebook, Amazon) as suppliers of various services in our Internet-dominated times.

As I write (April 2021), the Covid-19 pandemic rages. It starts being controlled through vaccination. Two countries, the US and the UK, have vaccinated a significant segment of their population. No wonder: they have devised and mass-produced the vaccines (Pfizer, AstraZeneca (Oxford), Moderna, Johnson & Johnson). That producers are also the predominant users, at least initially, should come as no surprise. Relevant figures are (April 22nd 2021), the UK has 42% of its population vaccinated, the US 40.6%, versus only 19.9% in France. The Anglo-Saxons have done twice better than we French, at this stage.

There is a direct correlation with Anglo-Saxon hegemony over the biomedical sciences. It dates back to the previous century: use of English as the publication language of articles included in PubMed—the catalog of relevant articles in biomedical journals published online by the NIH, the American National Institutes of Health—has gradually risen from 62.3% of the total number of indexed articles between 1967–1976, to 74.0% between 1977–1986, 83.4% between 1987–1996, and reached 89.3% in the period between 1997–2006 (unfortunately, I am unable to find and thus state more recent data). Turning to Nobel prizes in physiology or medicine, during the first 20 years of the twenty-first century, they were still awarded to Anglo-Saxon scientists, American and English, no fewer than 29 times—the lion's share.

My age, upon first arrival in the US (1962) and numerous subsequent stays and visits, gave me the privilege of becoming at least acquainted and frequently a good friend of a number of immigrants, each of whom has strongly benefited American science. I'll turn now to the topic of their determinant contribution to the world science. For that, I will draw merely upon my personal acquaintances.

The first contingent I'll mention is that of the Jewish immigrants, who fled Europe during World War II or in its immediate aftermath. Of the six who come immediately to my mind, two transited through Cuba before admittance in the US, Ernest L. Eliel (1921–2008) and Earl Peters (1927–2017). With the coming of the Nazis to power in 1933, Ernest left Germany and moved to Scotland, then Canada, then Cuba. He received his B.S. from the University of Havana in 1946. Why Cuba? One has to recall that Jewish refugees were undesirable on many countries, including the US, during most of the 1940s. Cuba admitted them, though, and was near the USA— hence it became an elected destination to Ernest.

The same was true in Earl's story. He fled Germany in 1939 with his parents, Stanley and Pauline. Earl Peters and his parents were on the ill-fated not to say doomed ship, the *St. Louis* that went from Hamburg to Cuba in May 1939, with 937 passengers, most of them Jewish refugees from Nazi persecution in Germany. The Cuban, Canadian and American governments all refused to accept the foreign refugees and the *St. Louis* returned to Europe with 908 passengers. Luckily, Earl got sick while in Havana and was allowed to disembark and be hospitalized there. It may have saved his life. After living in Cuba for two years, the family arrived in Buffalo, NY in 1941 as refugees.

Tom Eisner's (1929–2011) escape from Nazi Germany was a little different: Uruguay rather than Cuba. His parents, Hans E. Eisner (1892–1983) and Margarete Heil Eisner, fled Germany with their two children in April 1933, the year when Hitler became Chancellor. Hans E. Eisner was among the last graduate students of Nobel Chemistry Laureate Fritz Haber and worked with Haber at the Kaiser Wilhelm Institute for Physical Chemistry and Electrochemistry in Berlin. They went first to Barcelona, but decided to leave Europe and settle in Uruguay. To improve the education of their children, the Eisners immigrated to the United States.

What a bonanza: Ernest would become a professor of chemistry at the University of North Carolina in Chapel Hill and president of the American Chemical Society; Earl would become executive director of the Department of Chemistry at Cornell; and Tom, together (of course) with Jerry (Meinwald, 1927–2018), a colleague of his at Cornell, would pioneer the brand-new field of chemical ecology.

Another two I also came into close association with were Carl Djerassi (1923–2015) and Kurt Mislow (1923–2017). The former but not the latter also had a trajectory that included a Latin American country: from a family having emigrated from Bulgaria, Carl was born in Vienna, the son of Samuel Djerassi, a dermatologist and sexual health specialist, and Alice Friedmann, a dentist and physician. He and his mother escaped from Vienna at the time of the Anschluss when the Nazis took over in 1938. They ended up in the US the following year. His father did not make the journey until 1949. In the US, Carl was educated at Kenyon College in Ohio, and received his Ph.D. from the University of Wisconsin in Madison in 1945. After four years with the Ciba pharmaceutical company in New Jersey, he joined Syntex in Mexico in 1949 as associate director of medical research. He worked there on the chemistry of steroids, obtained from a natural source, a Mexican wild yam, and was thus instrumental in developing the birth-control pill.

Roald Hoffmann (b1937) became a close personal friend and I am very proud of our joint papers. His experience during World War II was most dramatic. He was born in Złoczów,in the Ukrainian part of then Poland, to a Polish-Jewish family. His first name honors the Norwegian explorer Roald Amundsen. His parents were Clara (Rosen), a teacher, and Hillel Safran, a civil engineer. After Germany invaded Poland and occupied the town, his family was placed in a labor camp where his father, who was familiar with much of the local infrastructure, was a valued prisoner. As the situation grew more dangerous, with prisoners being transferred to extermination camps, the family bribed guards to allow an escape. They arranged with a Ukrainian neighbor named Mykola Dyuk for Hoffmann, his mother, two uncles and an aunt to hide in the attic and a storeroom of the local schoolhouse, where they remained for eighteen months, from January 1943 to June 1944, while Hoffmann was aged 5 to 7. His father remained at the labor camp but was able to occasionally visit until he was tortured and killed by the Germans for plotting to escape. Roald and his mother migrated to the United States on the troop carrier *Ernie Pyle* in 1949. She had remarried Paul Hoffmann in Crakow: he was a kind and gentle father to Roald until his death in 1981, only two months prior to the announcement of Roald's being awarded a Nobel Prize.

Roald is the proverbial Renaissance man: in addition to an overwhelming contribution to chemistry, he is a poet, a playwright and an essayist of distinction. I pride myself on having translated into French a few of his poems. He and I wrote together four well-received essays on topics at the juncture of chemistry, history and philosophy. Roald introduced me to Oliver Sacks (1933–2015), whom I was thus fortunate to know personally.

Which brings up my connections with the five other above-mentioned immigrants to the US of Jewish origins. Ernest Eliel? We first met during the early Sixties, at some conference in Europe. He was prestigious to me, from his contributions to conformational analysis, then at the fore. He proudly told me that Cologne had been his hometown, prior to his forced emigration to the US. I met him again, during the late Sixties, at Princeton. He was spending a sabbatical in Kurt Mislow's group. I was an assistant professor. At least in my eyes, Ernest paled in comparison to Kurt.

Later on, towards the end of the century, Valerie and I had our American home in Pinehurst, North Carolina—not far from Chapel Hill and the University of North Carolina campus, where by then Ernest was a professor. My most vivid recollections, though, are not of Ernest but of his wife Eva Schwarz Eliel (1924–2013). They had married in 1949. Born in Hamburg, Germany, Eva came to the United States in 1937. She graduated from Hunter College in New York City and then worked for *Time* magazine. For many years, and that is my recollection, Eva had a daily morning show on the WUNC radio station, with a classical music program: she always sounded confident and assured, a both comforting and companionable voice.

I first met Earl Peters during my first visiting professorship at Cornell. He was then Executive Director in the Chemistry Department. He was extremely congenial and helpful to me, and went out of his way to make sure i would not feel lonely and would meet colleagues who would also be friendly. He and his wife Harriet Peters (b 1936), who worked in the Department of Management, became good friends. Once, in the late Eighties, when Earl made a trip to Europe he came to visit us in our house in St. Rémy-lès-Chevreuse. I vividly recall, while the RER suburban train carrying us went through Bures-sur-Yvette, followed by Gif-sur-Yvette, Earl asking in a false naive manner, "who is Yvette?" (Yvette is the name of a small stream).

Tom Eisner? He and his wife Maria became good friends during my first stay in Ithaca. Was it then or during a later visiting professorship, I audited his class in chemical ecology? He would spend his summers in the American Southwest, chasing and collecting insects, coleopters mostly, photographing them. He thus enjoyed a time in the limelight for his movies of the bombardier beetle in action. I own a lasting remembrance of Tom, when asked for a recommendation of a forgotten book that deserved to have become a classic of science, he pointed me to Kenneth D. Roeder's *Nerve Cells and Insect Behavior*, Harvard, 1967.

Carl Djerassi? Quite a personality. An authoritarian, very intense man. Very wealthy from his former position at Syntex and the heyday of steroidal hormones, he used his money to good purpose: an art collection of works by Paul Klee (1879–1940) donated to the Museum in San Francisco, and his former ranch overlooking the Pacific turned into a Foundation for artists and writers—similar to Les Treilles, mentioned in an earlier chapter. I stayed there I believe three memorable and productive times.

I first saw him at a GECO conference in Sarlat, in the mid-Sixties. As a chemist, Djerassi was one of the stars at Stanford. He campaigned eagerly for a Nobel prize, that he did not succeed though in getting. I owe him the already mentioned invitation to visit the Swedish Polar Station in Abisko. I was very fond of his (last) wife, Diane Middlebrook (1939–2007), also a professor at Stanford. She was a darling, a joy to talk to. I visited their flat in London, in the handsome Little Venice area, a couple times. As is well known, Carl made a name for himself in literature—in part as a way to court Diane. Indeed, I took part in a Festschrift symposium in his honor at the University of Dortmund. My contribution dealt (teasingly) with the academic novel—the book from that sympisium appeared in 2012. Carl did not rise to the bait and complimented me instead. Together with Roald Hoffmann he wrote the play *Oxygen*, at a time when Tom Stoppard (b 1937) had rendered popular such science-historical plays. He and I wrote together NO, a play meant for pupils in the schoolroom.

Kurt Mislow? I have mentioned him already a number of times in this account. We first met during an ACS conference in New York City, in 1963 I believe, where I had accompanied Paul Schleyer, who made the introduction. Kurt was then still at NYU. I remember being impressed with his stamina as a New Yorker, capable that he was of a loud whistle to hail a taxi by simply raising fingers to his lips. Our next meeting was at the Bürgenstock conference in stereochemistry in 1966. Kurt was proudly showing everyone photographs of Jacqueline Ford, a medical doctor and internist, his second wife to be. In the fall of 1966, I moved to Princeton as a junior faculty member. Shortly afterwards, Kurt replaced Walter Kauzmann (1916–2009) as the chairman of the Department. Kurt was an outstanding scientist, a deep thinker and an inspiration.

The second contingent I'll mention is that of the Hungarian immigrants who fled their country for the US following the Budapest uprising and its subsequent crushing by the Red Army in 1956. I had the good fortune of becoming acquainted and a friend of three of them, who all became stars of American and world science, George A. Olah (1927–2017), Gabor Somorjai (b 1935) and Peter J. Stang (b 1941). Given my acquaintance from childhood with the Hungarian language, I could detect in each of their accents a lingering reminiscence of their mother tongue.

I became first acquainted with George Olah in the same way as with Kurt Mislow, having accompanied Paul Schleyer to a meeting in New York. He was then chairman of the Department of Chemistry at Case-Western Reserve in Cleveland, Ohio. George Olah had a career as a chemist, from Budapest to (1956) Ontario and (1964) Massachusetts, followed by (1965) Ohio and finally (1977) California.

Born into a middle-class Budapest family, their apartment across from the Opera House, Olah was given a well-rounded education, in a secondary school run by the Roman Catholic Piarist Fathers (similar in that to my father's education), with eight years of compulsory Latin and German and elective French for four years. He was also tutored in French and English. His schooling was likewise comprehensive in maths and the sciences.

After graduation in 1945, he entered the Technical University of Budapest (again like my father), where the curriculum was demanding and the exams sieved out all but the top students. He started research in the Organic Chemistry Institute, at the Technical University, continuing it at the Academy of Sciences. He then became part of a diaspora of well-trained and very bright Hungarian scientists who left their country, in November and December 1956, in the aftermath of the failed uprising against Soviet rule.

After fleeing Hungary in 1956, Olah joined the Sarnia, Ontario laboratory of the Dow Chemical Company, from which he transferred in 1964 to another Dow lab, in Framingham, Massachusetts, under enlightened steering by Fred W. McLafferty (b 1923). This was the time when American industrial laboratories, such as Bell Labs, DuPont de Nemours, Eastman Kodak, Xerox, Union Carbide or Dow, carried out pure science, worthy of the best research universities. The context of Olah's scientific contributions, on positively charged intermediates, known as carbocations, also included the instrumental revolution, that brought powerful new spectroscopies, nuclear magnetic resonance (nmr) predominantly, into chemical laboratories. Olah had the lucidity and foresight to jump early onto the nmr bandwagon. George Olah's career embraced this glorious period in organic chemistry, the second half of the twentieth century. He was lucky to reach North America when the Cold War and Sputnik had spawned a massive expansion of research universities in the United States and of their public support. In 1965, Olah indeed moved to Western Reserve University in Cleveland; in 1974 to the University of Southern California (USC) in Los Angeles.

Outward but given to outbursts; most congenial one-on-one, Olah was prone to writing scathing referee reports and tended to be aggressive in public discussions. His wife, Judith Agnes (b 1928), was a teenage love. They were married and had their first child before escaping from Hungary in 1956. When he espoused academic life in 1965 in Cleveland, she not only entered his research group, she helped to run it and to maintain a congenial atmosphere, freeing him from chores, such as ordering supplies. Because of her own involvement in research and since he was a workaholic, he could confide in her and rely on her sympathetic and knowledgeable attention.

A gripping part of their story is the brief account of the two major health crises George Olah overcame. In 1979, he was felled by a mysterious ailment. It was diagnosed after he and Judith, in desperation, rushed to the Mayo Clinic in Rochester, Minnesota: pamphigus is an orphan immune disease—probably caused in his case by traces of penicillamine lingering in the USC building where he had his office, due to earlier, wartime research on penicillin. A piece of detection, made easier by a colleague in a nearby office suffering from the same ailment, but illustrating Olah's skill in problem-solving. And, in 1982, Olah contracted stomach cancer, from which he was operated on successfully. The chemical community gasped in relief, it was obvious that Olah would be awarded a Nobel prize, if only he survived that illness.

His contributions to organic chemistry stem from his use of superacids, with acidities billions of times stronger than concentrated sulfuric acid, that stabilize and make apparent otherwise elusive species. He shone by his talents as an experimentalist and as an organizer, deep incisive thinking and the ability at innovative conceptualization.

A key time in his career was the classical-non classical ion bitter controversy of the Sixties. Olah and his friends were on the winning side. Yet, this renewed feud of the Ancients (Herbert C. Brown (1912–2004), a born, sometimes disingenuous debater) and the Moderns (Saul Winstein, 1912–1969) caused in 1970 the downfall of physical organic chemistry: its funding by American Federal agencies abruptly dried-up.

What made Olah a successful scientist, a Nobel prizewinner? Reading and writing. Chemistry not only is cumulative, it is also very much textual, in the form sometimes of a palimpsest. Olah relished reading for long hours in libraries, which is how he found his career's call, a direction he would hold to with remarkable persistence:" after having read about organic fluorine compounds, I became interested in them." He prided himself in continuing into the computer age to write longhand.

The young Olah, rebelling against his old-fashioned Hungarian mentor Géza Zemplén (1883–1956), interested himself in reaction mechanisms and intermediates. Not only metaphorically, physically as well: he had to stake out a workplace of his own, "at the top floor of the chemistry building, overlooking the Danube and badly damaged during the war, was an open balcony used to store chemicals." Then, decades later, the mature author found wealthy donors to build him at USC a whole institute and library devoted to his research.

Olah vowed that his Nobel prize (1994) would not end his career, as is all too often the case. He prevailed and went on creating and publishing chemistry of lasting import. Concerned about the current reliance on fossil fuels and the ensuing global warming, Olah worked out a methanol economy, where methanol fuel would be made from recovered carbon dioxide.

Two fellow-chemists who also left Hungary in 1956 and who became likewise leaders in American chemistry were indeed Gabor A. Somorjai, a Wolf Prize laureate, whose life had been saved by Raoul Wallenberg (1912-?) in 1944; and Peter Stang, a Priestley medalist of the American Chemical Society.

My first meeting with the Somorjais, both Gabor and his wife Judith K. was at a Bürgenstock conference, in 1989. Was it because he considered me as a fellow-Hungarian or because of our then ongoing work about catalysis or organic reactions on the surfaces of inorganic solids? In any case, he invited me right away to visit his Institute at Berkeley.

Thus, Valerie and I spent one month there during the following year, 1990. I found Gabor to be a most considerate host, for instance having me accompany him to a faculty meeting of the chemistry department. His Institute was located at the top of the campus. If great for tranquility this was not inducive to encounters with fellow chemists, whose labs were down at the center of the campus.

Somorjai had also been a student at the Technical University in Budapest, in chemical engineering, starting in 1953. He took part in the 1956 uprising. "When the Russians began arresting people, I decided to leave the country. I was four months from getting my diploma. I hid my machine gun in my organic lab locker and headed for the border." He took with him his sister Marietta, and, after consulting with her parents, his girlfriend (and later wife), Judith Kaldor. He was 21, she was 18. Gabor and Judith had to walk into Austria in order to escape from Hungary: "with soldiers guarding the final crossing, we had to leave the train 50 kilometers from the border. The local people hid us during the day, and we walked for four nights to the border, where guides led us to an area where we could cross safely. I remember it was late November 1956, cold and swampy."

They went on to Vienna, where they encountered fellow Hungarian émigré Cornelius Tobias, brother of Charles Tobias (1920–1996) who was on the faculty at Berkeley. Somorjai traveled to the United States with Judith, where they had been sponsored by refugee agencies. Their first home in the U.S. was Camp Kilmer, NJ. Charles Tobias arranged in 1957 for Gabor to become accepted as a graduate student at Berkeley, with a stipend of $800. Gabor and Judith were married in the fall of 1957. For their honeymoon they took a bus to Lake Tahoe over the Labor Day weekend.

Almost his whole career and their whole life were spent at the University of California. In 1960, after getting his Ph. D., Gabor joined the research staff at IBM in Yorktown Heights, NY. In 1964, he decided to quit IBM and return to Berkeley as a professor and to continue studying surface chemistry.

A quarter of a century later, when I visited, his Institute was very impressive, from both its equipment and an outstanding staff. I found Gabor to be a little cold and stiff—probably because of the busy schedule he kept.

Now to Peter Stang, whom I have known for the last 56 years! We met in the fall of 1966, he had just arrived at Princeton to post-doc with Paul Schleyer and I was a young faculty member with my own research group. I have recounted earlier in this book how Peter convinced me to turn my classes on the determination of structures by physical methods into a book that we wrote jointly. Entitled *Organic Spectroscopy* it was published in 1971 by Harper and Row, in New York.

Peter Stang was a teenager during the 1956 Hungarian uprising against Soviet rule. "The revolution was successful for a few days," he recalls. "Then one night my father, who spoke English, was listening to the BBC on the radio. They reported that three Russian divisions had crossed the border into Hungary. My father announced: 'Tomorrow, we leave.'"

That night the entire family—Stang, his parents, and two sisters—packed a few things. They caught a westbound train in the morning and disembarked that night to start a 25-mile trek—similar to the Olahs's—across the Hungarian/Austrian border on foot.

Stang's scientific prowess earned him a spot in DePaul University's freshman class. After graduating, Stang earned two National Institutes of Health fellowships that took him first to the University of California (Berkeley) for his Ph.D. with Andrew D. Streitwieser and then to Princeton University for his postdoctoral work.

When at Princeton, every night Peter found correspondents to talk to, all over, through short-wave radio. This was in the late Sixties, when East Europe was still under Communist rule. One of my recollections is of Peter's excitement when the Yugoslav politician Milovan Djilas (1911–1995) visited Princeton, I believe they met. Djilas was one of the best-known and most prominent dissidents in the Eastern Block.

Peter was appointed afterwards to a professorship at the University of Utah. Over the years, we have maintained our friendship. He has done superb work in chemistry: a kind of Erector set, with molecules as the elements—in so doing, Peter has carved out an entirely new chapter of chemistry. It has brought him the highest and well-earned honors. Yet, Peter remains amazingly modest about his merits—because he admires the achievements of others and is not self-absorbed, is my hunch.

The US became the foremost economy as a consequence of the two world wars. Immigration to the US was another consequence and it has benefited American science and universities as I have just documented with these personal examples. Will the top universities, the American and Oxbridge, be able to keep it up? Yes, I strongly believe. In recent decades, they have started to diversify—perhaps more successfully with inclusion of women on their faculties more than with African-Americans and Latinos; three of the seven appointments in chemistry at Caltech I mentioned earlier went to women.

The next challenge, a golden opportunity as well, will be to tap the big reservoir of talents South of the Rio Grande—the whole Latin-American sub-continent. Success will depend on maintaining the integrity of their faculty appointment process, of which tenure is a key element.

Thus, I trust others to carry the torch lighting our collective way to advancement of knowledge.

Envoi

To be lucky is essential. To be endowed with curiosity is at least equally important. I followed my nose wherever it led me. Quite consciously: the very first words in *La parole des choses*, also my very first book on popularization of chemistry, are *je hume* (I sniff). Yes, I am also a gourmet; and I love to cook.

This helps to explain my somewhat ecumenical interests. Together with a couple other ingredients, an appetite for reading books of every kind, that my parents fostered in childhood; and the encyclopedic program of study in NSE, the *classe préparatoire* I enrolled in as a teenager. Plus the need to become familiar with a topic that has struck my attention—as Simone Weil wrote more or less (I am quoting from memory), "to be attentive is a gift from God."

To convey how ecumenical indeed my interests are, my office in Liège had dozens of binders on its shelves. I stored in them reprints of articles from a large number of fields. Everyday, I sent out reprint requests to colleagues and received quite a few. I left all those binders behind when I left Liège for Palaiseau and I don't know their destiny—a trashcan probably.

Lucky me! To be a member of the worldwide scientific community. To be called upon to lecture and present our work at the four corners of the planet.

A significant reward was to experience all these amazing landscapes and townscapes, such as hiking on the Tokaïdo trail in Japan; the sparkling townscape of Hong-Kong during a flyover at night; Upstate New York and its rolling hills; the Outer Banks in North Carolina; Yosemite; the wildernesses in Wyoming; the city of Ouro Preto in Brazil; also in Brazil, Serra dos Órgãos, near Teresópolis; the Austrian province of Carinthia; Norway and New Zealand and Siberia, all three graced with splendid mountain ranges; ancient small cities such as, in Britain, York, Chester and Norwich, in Italy, Lucca and Urbino; in France, Arles and Uzès and the southern *bastides;* large vibrant cities, such as in Europe Amsterdam and Barcelona; places of reverence, the Greek temples in Agrigento, the Priory of Serrabone in the Roussillon

. . .

I was graced with a charmed life.

Sénergues, Christmas Day, 2020.

P. Laszlo, *A Life and Career in Chemistry*,
https://doi.org/10.1007/978-3-030-82393-1

Index

Printed in the United States
by Baker & Taylor Publisher Services